AGILE IN HARDWARE, ENGINEERING, CONSTRUCTION, AND LINEAR ENVIRONMENTS

A Guide for Applying Agile Thinking and Practices

PHILL AKINWALE, OPM3, PMP, PMI-ACP, PSM, CSM, PSPO, SPS

PUBLISHER

Praizion Media

Praizion Media
Real World Project Management Training Solutions

AGILE IN HARDWARE, ENGINEERING, CONSTRUCTION AND LINEAR ENVIRONMENTS
Published by Praizion Media
P.O Box 22241, Mesa, AZ 85277
E-mail: info@praizion.com
www.praizion.com

Author
Phillip Akinwale, MSc, OPM3, PMP, PMI -RMP, PMI-SP, PMI-ACP, PSM, CSM, PSPO, SPS

9 781934 579763

ISBN 978-1-934579-76-3

Printed in the United States of America

Table of Contents

Chapter 1: The Essence Of Agile Thinking: Beyond Tools And Frameworks 17

Agile In Business Case Study #1 22

Chapter 2: Universality Of Agile Principles: Bridging Industries 24

Chapter 3: Agile Manifesto Values: A Universal Compass For Success 38

Agile In Business Case Study #2 45

Chapter 4: Self-Organization And Inspection: Nurturing Agile Excellence 47

Chapter 5: Agile Beyond Frameworks: Embracing The Agile Mindset 57

Chapter 6: Network Implementation With Agile: Bridging The Digital Divide 64

Chapter 7: Creative Value Delivery: Unconventional Approaches To Agile Success 73

Chapter 8: The Customer's Perspective: Value From Their Eyes 86

Chapter 9: Agile Challenges And Solutions: Thriving Amid Complexity 91

Chapter 10: Continuous Innovation And Adaptation: The Agile Imperative 98

Chapter 11: Agile Success Stories: Real-World Examples Of Transformation 106

Construction Software - Case Study A 107

Oil And Gas - Case Study B 108

Aeronautics - Case Study C 110

Biotech - Case Study D 112

Aerospace, Defense & Security, Transportation - Case Study E 114

Hardware: Network Installation Reinvented Hypothetical Case Study 116

Manufacturing: Lean And Agile Production Hypothetical Case Study 117

Construction Hypothetcal Case Study 119

Chapter 12: Final Thoughts 121

Table of figures

Figure 1: Roadmap to Agility #1 15
Figure 2: Roadmap to Agility #2 16

INTRODUCTION

"

"It is not the one who uses the framework that is agile. It is the one who employs agile thinking by putting humans first and pivoting to accommodate the constant of change."

Phill Akinwale

Agile is a MINDSET! NOT a methodology or framework! Let's make that clear immediately.

Agile, a term that has permeated the lexicon of industries far and wide, often finds itself ensnared in a web of misunderstanding. Many, with earnest intentions, seek to grasp its essence, yet find themselves entangled in the trappings of methods, frameworks, processes, and tools.

“Agile is not an end in itself; it is a means to an end,” a profound realization that beckons us to venture deeper, to seek the true nature of agility.

When we embark on this intellectual odyssey, we uncover that Agile is not a mere collection of practices to be implemented; it is a mindset, a philosophical approach to work and life. “Agility is the ability to adapt and respond to change... agile organizations view change as an opportunity, not a threat,” a perspective that transcends industries and sectors, including the linear realms of construction, engineering, and hardware.

Detractors may argue, with fervor and conviction, that the incremental delivery of value, a cornerstone of Agile philosophy, is an impossibility in these industries. They point to the tangible, the concrete, the steel, and proclaim, “Here, Agile cannot tread.” Yet, this argument, while rooted in the realities of these industries, misses the mark. “Value is in the eye of the beholder,” and in this, we find the key to unlocking Agile’s potential across all domains.

Value, in its multifaceted glory, is not confined to the delivery of a product or a component. It is time reclaimed, costs saved, satisfaction achieved, awareness heightened. It is both tangible and intangible, monetary and non-monetary. “The main goal of Agile is not delivering a deliverable, but delivering value,” a paradigm shift that challenges us to rethink our approach to work.

To embody agility, to be truly agile, our focus must shift from the delivery of parts of a solution to the relentless pursuit of value delivery. This is the crucible of Agile transformation, a journey fraught with challenges, yet rich with rewards. “Agility requires a shift in mindset, not just a change in practices," a truth that becomes self-evident as we navigate the complexities of applying Agile principles in hardware, construction, and engineering.

The question then transforms from, "How can we deliver a part of the solution?" to "How can we deliver value at the end of each iteration, regardless of the industry?" In instances where a working solution is an impossibility, we must not retreat, declaring Agile unfit for our domain. Instead, we must persevere, delivering value in other forms—schematics, drawings, information, progress updates—demonstrating to the client that we are on the right path.

"Agile is a journey, not a destination," a journey that requires us to approach each project with a mindset of value delivery and continuous improvement. It is not about adorning ourselves with the label of Agile; it is about enhancing our way of working, being as agile as possible, delivering unparalleled value and benefits, and adopting a fluid, dynamic approach that opens up a world of possibilities.

"The customer is the ultimate arbiter of value," and in this, we find our true north. Our relentless pursuit of value, coupled with an unwavering commitment to seeking the customer's feedback, ensures that we are not just on a path, but on the right path. Agile, in its truest form, is a mindset, a way of thinking and being that, when embraced, has the power to transform not just our projects, but our very lives.

Mindsets drive behavior of individuals and teams on any project in any industry. So therefore, Agile thinking can be applied to any industry, including hardware through the execution by individuals employing agile-mindedness. Agile, at its core, transcends the boundaries of a mere methodology; it embodies a philosophical echelon, a paradigm of thought that hinges on adaptability, collaboration, and iterative progression. This mindset, once embraced, becomes the compass that navigates the behavior of individuals and collectives across the diverse landscapes of projects, irrespective of the industry.

As an Agile coach with extensive experience across a myriad of industries including manufacturing, engineering, construction, architecture, oil and gas, healthcare, and more, I can assertively state that the Agile framework holds a universal applicability. My co-trainer Roy and I have had the privilege of guiding both private and government entities through the Agile transformation journey. Our experiences have unveiled that with a balanced blend of patience, understanding, experimentation, and education, any environment can morph into a fertile ground for Agile practices to flourish.

A notable testament to Agile's versatility is its successful application in the government sector, particularly illustrated by the Virtual Case File project undertaken by the FBI. This project marked a pivotal shift from a predictive to an Agile approach, embodying the principles of iterative and incremental progress. Through the disciplined adoption of Agile practices via the Scrum framework, the FBI managed to significantly revamp their operational ethos. This transition facilitated the actualization of their project goals, enabling them to construct the necessary solutions effectively.

The overarching narrative here underscores Agile's profound capacity to be tailored to any project milieu, fostering a culture of continuous improvement, collaboration, and value-driven delivery. Whether in the bustling corridors of a manufacturing facility or the stringent realms of government operations, Agile unveils a pathway to enhanced project efficacy and adaptability. Through our training endeavors across diverse fields, Roy and I have witnessed the transformative power of Agile, affirming its stance as a universally adaptable approach.

The doctrine of Agile isn't confined to the silo of software development; its essence is universal, its principles versatile. Hence, its application flourishes beyond the digital realm, extending its relevance to tangible, hardware-centric industries. The conduit for this transference of Agile philosophy lies in the minds and actions of individuals who, with a penchant for agile-mindedness, become the harbingers of a flexible, responsive ethos within their teams and projects.

As these agile-minded individuals navigate the intricacies of project execution, they carry with them the torch of Agile thinking, illuminating pathways of iterative problem-solving, feedback-driven improvement, and collaborative endeavor. Through their adept execution, the Agile mindset permeates the fabric of the project, nurturing a culture of continuous evolution and reciprocal growth. Thus, Agile transcends the theoretical and melds with the practical, its essence echoed in the rhythmic cadence of iterative development, whether in realms of code or the tactile world of hardware. Through the lens of Agile, every task becomes an opportunity for reflection, adaptation, and relentless pursuit of excellence, fostering a symbiotic evolution between the project and its architects.

Agile has revolutionized the way projects are managed and executed across various domains, transcending its origins in software development. It is a common misconception that the essence of Agile is encapsulated in the tools, frameworks, or practices that are commonly associated with it, such as Scrum, Kanban, or the use of project management tools like Jira,

Confluence, or VersionOne. However, the true spirit of Agile lies far beyond these tangible elements.

Agile Thinking and Behavior: The Core of Agility

Agile is fundamentally a mindset, a philosophy, a posture, a way of thinking and behaving that emphasizes flexibility, collaboration, and customer centricity. It is about embracing change, even late in the development process, and delivering value to the customer incrementally and iteratively. This mindset can be applied universally, regardless of the nature of the project, be it software, hardware, or any other domain.

Iterative Planning and Incremental Delivery

One of the key hallmarks of Agile is iterative planning and incremental delivery. This means that work is broken down into small, manageable chunks, and delivered in iterations, allowing for frequent reassessment and adaptation of plans. Each increment adds value and is functional or shippable, ensuring that the project is always moving forward and that value is continuously delivered to the customer.

Agile Mindset: Individuals and Interactions Over Processes and Tools

The Agile Manifesto emphasizes the importance of individuals and interactions over processes and tools, working product over comprehensive documentation, customer collaboration over contract negotiation, and responding to change over following a plan. This highlights the human element of Agile, putting the focus on collaboration, communication, and adaptability.

Self-Management: The Agile Team Environment

Agile promotes a self-led, self-managed, and self-organized team environment. This empowers team members, giving them the autonomy to make decisions and take ownership of their work. It fosters a sense of responsibility and accountability, leading to higher engagement and productivity.

Frequent Inspection at All Levels

Agility requires frequent inspection at various levels, including the individual and team levels. This ensures that the project is on track, that issues are identified and addressed promptly, and that the team is continuously learning and improving. Inspection can happen at any point along the iteration, providing real-time feedback and allowing for immediate action.

Beyond Frameworks and Tools: Applying Agile Thinking

Agile is not prescriptive; there is no set number of practices or principles that must be applied to be considered Agile. It is about applying Agile thinking, behaviors, and ultimately practices in a way that suits the unique needs of the project and the team. This may involve combining multiple frameworks or ideas to create a unique blend that works best for the specific context.

Agile in Practice: A Network Implementation Example

Consider a network implementation across a large company campus. Applying an Agile approach would involve thorough planning, using an appropriate framework, and ensuring that Agile practices, principles, and values are upheld. It requires a team charter, breaking work down into sprints or iterations, and frequent inspection and adaptation.

Even in contexts where it may seem challenging to deliver a chunk of functionality to the customer in each iteration, such as in construction or hardware projects, it is important to seek creative ways to deliver value continuously. This could involve providing schematics, drawings, or regular updates to the customer, ensuring that the customer's perception of value is always considered.

Roadmap to Agility: 10 Steps to Greater Agility

Step 1: Grasp Agile Fundamentals

- 1.1 Train All Levels: Ensure comprehensive Agile training for leadership and teams.
- 1.2 Adapt Agile Manifesto: Tailor the Agile Manifesto's principles to hardware and construction contexts.

Step 2: Evaluate Current Practices

- 2.1 Audit Processes: Review existing project and development workflows for potential enhancements.
- 2.2 Recognize Constraints: Identify and understand the specific challenges in hardware and construction, such as regulatory standards and material lead times.

Step 3: Cultivate an Agile Mindset

- 3.1 Promote Collaboration: Foster a collaborative culture across departments.
- 3.2 Welcome Change: Encourage adaptability and ongoing improvement.

Step 4: Apply Agile Techniques

- 4.1 Start with a Pilot: Initiate Agile practices with a small project or team.
- 4.2 Choose and Adapt Frameworks: Select and tailor an Agile framework like Scrum or Kanban to your project's needs.
- 4.3 Prioritize Customer Interaction: Maintain regular communication with customers for feedback and alignment.

Step 5: Boost Communication and Teamwork

- 5.1 Implement Visual Tools: Use visual aids for better project transparency and communication.
- 5.2 Build Cross-Functional Teams: Form teams with diverse departmental expertise.

Step 6: Adopt Iterative Development

- 6.1 Segment the Project: Break the project into smaller, iterative parts.
- 6.2 Prioritize Value: Focus on tasks that offer the most customer and project value.

Step 7: Create Feedback Loops

- 7.1 Regular Reviews: Conduct reviews after each iteration for progress assessment and feedback.
- 7.2 Pursue Continuous Improvement: Utilize feedback for ongoing process and product enhancement.

Step 8: Expand Agile Practices

- 8.1 Widen Agile Adoption: Gradually introduce Agile practices to more teams and projects.
- 8.2 Continuously Adapt: Keep evolving your Agile practices to meet the unique demands of hardware and construction projects.

Step 9: Track and Celebrate Progress

- 9.1 Set Success Metrics: Define clear metrics to gauge your Agile implementation's success.
- 9.2 Acknowledge Success: Celebrate milestones to sustain motivation and momentum.

Step 10: Disseminate Knowledge

- 10.1 Document Learnings: Record lessons learned throughout the Agile journey.
- 10.2 Share Insights: Promote a learning culture by sharing experiences across teams and projects.

Conclusion

In conclusion, Agile is not defined by the frameworks, tools, or methodologies used; it is a mindset, a way of thinking and behaving that emphasizes flexibility, collaboration, and customer centricity. It requires a shift in mindset, away from traditional, rigid ways of working, towards a more adaptive, collaborative, and customer-focused approach. By embracing Agile thinking and behaviors, and applying them consistently across all aspects of a project, teams can truly embody the spirit of Agile, leading to more successful, adaptive, and value-driven projects.

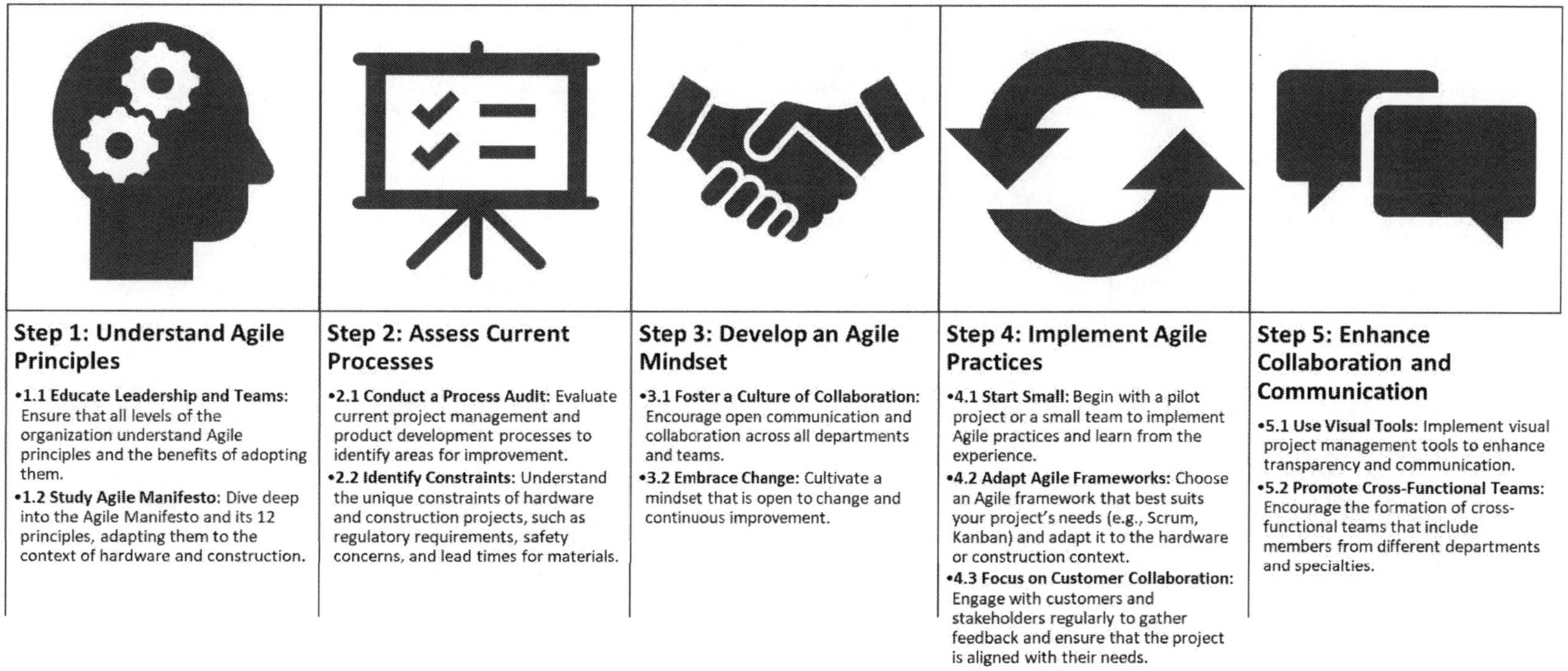

Figure 1: Roadmap to Agility #1

Step 6: Implement Iterative Development	Step 7: Establish Feedback Loops	Step 8: Scale Agile Practices	Step 9: Measure and Celebrate Success	Step 10: Share Knowledge and Learnings
• **6.1 Break Down the Project**: Divide the project into smaller, manageable parts that can be completed in short iterations. • **6.2 Focus on Delivering Value**: Prioritize tasks based on their value to the customer and the project.	• **7.1 Conduct Regular Reviews**: Hold regular reviews at the end of each iteration to assess progress and gather feedback. • **7.2 Implement Continuous Improvement**: Use feedback to continuously improve processes, practices, and the product itself.	• **8.1 Expand to Other Teams and Projects**: As you gain experience and confidence in Agile practices, gradually expand them to other teams and projects. • **8.2 Adapt and Evolve**: Continuously adapt and evolve your Agile practices to suit the unique needs of hardware and construction projects.	• **9.1 Establish Metrics**: Define clear metrics to measure the success of your Agile implementation. • **9.2 Celebrate Achievements**: Celebrate milestones and achievements to maintain momentum and motivation.	• **10.1 Document Lessons Learned**: Document lessons learned throughout the Agile implementation process. • **10.2 Share Knowledge**: Share knowledge and experiences across teams and projects to foster a culture of learning and continuous improvement.

Figure 2: Roadmap to Agility #2

Chapter 1: The Essence of Agile Thinking: Beyond Tools and Frameworks

"

"People are what make companies agile, not approaches, methodologies or tools."

– Pedro Gaspar Fernandes, consultant and trainer

Introduction

Agile thinking represents a revolutionary shift in how we approach work, collaborate, and deliver value. Originating from the software development industry, the agile philosophy has transcended its initial domain, proving its applicability and effectiveness across various fields. Whether in the construction of towering skyscrapers, the meticulous implementation of hardware systems, or the innovative development of software applications, adopting an agile mindset is synonymous with embracing adaptability, fostering collaboration, and ensuring continuous improvement.

The Genesis of Agile Thinking

The agile philosophy was born out of a need for an alternative to the traditional, rigid project management approaches that dominated the software development industry in the late 20th century. Developers were seeking a more flexible, responsive way to address the rapidly changing requirements and fast-paced nature of the tech world. In 2001, a group of software developers gathered to discuss these challenges, resulting in the Agile Manifesto—a declaration of four fundamental values and twelve guiding principles that encapsulate the essence of agile thinking.

Steve Jobs, co-founder of Apple Inc., was a proponent of innovative thinking and adaptability, qualities that are at the heart of the agile mindset. He once said, "Innovation distinguishes between a leader and a follower." Agile thinking is about leading the way, navigating through uncertainty, and being prepared to pivot when necessary. It's about fostering a culture of continuous improvement and innovation, ensuring that every project is an opportunity to learn and grow.

Agile Across Industries

While agile thinking has its roots in software development, its principles are universally applicable, transcending industry boundaries. In construction, agile thinking translates to adaptability in project management, responsiveness to client needs, and a collaborative approach to problem-solving. In hardware implementation, it means staying ahead of technological advancements, anticipating changes, and ensuring that systems are scalable and flexible.

Bill Gates, co-founder of Microsoft, emphasized the importance of adaptability when he said, “Success today requires the agility and drive to constantly rethink, reinvigorate, react, and reinvent.” This statement encapsulates the essence of agile thinking, highlighting the need for constant reevaluation and readiness to adapt in the face of change.

Fostering a Culture of Collaboration

Agile thinking places a strong emphasis on collaboration and effective communication. It’s about breaking down silos, encouraging cross-functional teamwork, and ensuring that every voice is heard. Les Brown, a renowned motivational speaker, captured the spirit of this collaborative approach when he said, “Help others achieve their dreams and you will achieve yours.” In the context of agile thinking, helping others means working together, sharing knowledge, and contributing to the collective success of the project.

Embracing Change as an Opportunity

Change is an integral part of the agile mindset. It’s about viewing change not as a threat, but as an opportunity for improvement and growth. Michael Dell, founder of Dell Technologies, once stated, “Whether you’ve found success or are still looking for it, I know you’re capable of so much more than you think you are.” Agile thinking challenges us to push our boundaries, to see beyond the current state of affairs, and to embrace change as a catalyst for greater success.

In conclusion, agile thinking is a transformative mindset that goes beyond specific tools and frameworks. It’s about adopting a flexible, responsive approach to work, fostering a culture of collaboration, and viewing change as an opportunity for improvement. Whether in construction, hardware implementation, or software development, agile thinking sets the foundation for innovation and excellence. As we venture deeper into the subsequent sections, we will explore how this mindset can be applied across various industries, demonstrating that agility is not confined by industry constraints but is instead a versatile approach to achieving success.

Overview of Agile in Construction

The construction industry has traditionally been known for its rigid and linear approach to project management. However, in recent years, there has been a growing interest in adopting agile practices to improve efficiency, collaboration, and overall project success. For professionals, engineers, construction experts, architects, project managers, scrum masters, product owners, and anyone involved in the construction industry, understanding the principles and practices of agile can greatly enhance their ability to deliver projects on time and within budget.

Agile project management in construction involves breaking down large projects into smaller, manageable tasks that can be completed in short iterations. This allows for flexibility and adaptability in response to changing requirements or unforeseen challenges. By embracing an iterative approach, construction teams can continuously improve their processes, identify and address issues early on, and deliver high-quality projects.

Overview of Agile in Hardware

Agile practices in hardware development are also gaining traction in the construction industry. By applying agile engineering processes, teams can streamline the design and manufacturing of building components, leading to faster production cycles and improved product quality. With agile approaches for linear infrastructure projects, such as roads, railways, and pipelines, construction teams can effectively manage complex and interconnected tasks, ensuring efficient project delivery.

Agile supply chain management in construction emphasizes collaboration and communication among suppliers, contractors, and other stakeholders. By adopting agile principles, construction companies can improve coordination, reduce delays, and enhance overall project performance.

Agile implementation in hardware manufacturing focuses on optimizing production processes, reducing waste, and increasing product innovation. By embracing agile principles, manufacturers can respond quickly to changing customer demands and market trends, leading to greater customer satisfaction and improved competitiveness.

Agile principles can also be applied to engineering consulting, enabling consultants to deliver tailored solutions that meet client needs while remaining adaptable to evolving project requirements

Agile in Business Case Study #1

Construction: Centrus Energy Corp.

Centrus Energy Corp., a leading supplier of enriched uranium fuel, engaged in a cooperative research, development, and demonstration (RD&D) program with the U.S. Department of Energy (DOE) from June 2012 to April 2014. The program aimed to validate the technical readiness of the American Centrifuge, a next-generation U.S. uranium enrichment technology. The project took place in Piketon, Ohio, with a budget of $350 million.

Program Results

The RD&D program was successfully completed within budget and on schedule, meeting all technical milestones and performance indicators set by the DOE. The project involved operating a commercial demonstration cascade, achieving significant operational milestones, and demonstrating the centrifuge's manufacturing quality, operational reliability, and sustained production capabilities. The program engaged 169 companies from 28 states, providing employment to over 1,100 workers.

Agile Implementation

Despite not formally adopting Agile, the project team informally applied agile principles to overcome challenges such as tight schedules, limited funding, and stringent regulatory oversight. Centrus acted as both the owner and at-risk construction manager, utilizing time and material contracts for most of the engineering, design, and construction labor.

The agile approach, characterized by an integrated team mindset and a focus on flexibility and collaboration, resulted in a project that was completed safely, on schedule, under budget, and with high productivity levels. This approach required enthusiastic participation from all stakeholders and a shared vision to keep the project on track.

Conclusion

The Centrus Energy Corp. case study demonstrates the successful application of agile philosophies and practices in a construction project, highlighting the benefits of flexibility, collaboration, and an integrated team approach in achieving project success, even in highly regulated environments.

Chapter 1 Exercise: The Essence of Agile Thinking: Beyond Tools and Frameworks

1. **Reflect on Traditional vs Agile Mindset**: Describe a situation from your own experience or a hypothetical scenario where a traditional project management approach failed, and discuss how an Agile mindset could have made a difference.

    - *Guidelines*: Focus on aspects like flexibility, customer collaboration, and iterative progress. Highlight the limitations of a rigid, plan-centric approach and how Agile could bring in adaptability and customer-centricity.

2. **Agile Tools and Frameworks**: List out some popular Agile tools and frameworks. Discuss how they could potentially be misused if the underlying Agile principles are not understood or followed.

    - *Guidelines*: Emphasize the importance of understanding the Agile principles and mindset before diving into tools and frameworks. Discuss how misuse can lead to a superficial implementation that doesn't yield the true benefits of Agile.

3. **Personal Agile Journey**: Reflect on your own work habits and mindset. In what ways could you adopt a more Agile approach in your daily tasks and interactions?

    - *Guidelines*: Encourage introspection and self-awareness. Discuss the importance of being open to change, collaborating effectively, and focusing on delivering value.

Chapter 2: Universality of Agile Principles: Bridging Industries

> *"Culture is like a shadow: You cannot change it, but it changes all the time."*
> *– Niels Pflaeging, author and consultant*

Introduction

Agile principles, originally formulated within the realm of software development, possess a remarkable universality, extending their influence far beyond their initial domain. This chapter ventures into the ways in which these principles bridge diverse industries, ranging from the structured world of construction to the intricate realm of hardware development. We will unravel the essence of agility, showcasing how its guiding principles serve as potent tools for achieving excellence across various fields.

The Essence of Agility Across Domains

Agility is not confined by industry boundaries; it is a mindset, a set of guiding principles that transcend the limitations of specific methodologies or frameworks. Its universality lies in its foundational values: adaptability, collaboration, and customer-centricity. These values are the common thread that weaves through various industries, providing a robust framework for excellence.

Adaptability: The Heartbeat of Agility

Agility thrives in environments that embrace change. In the construction industry, this adaptability might manifest in the ability to modify building plans in response to evolving client needs. In the realm of hardware, it could translate to adjusting network configurations to seamlessly integrate new technologies.

Collaboration: Uniting Diverse Talents

Agile principles champion the power of teamwork, open communication, and a focus on individuals and interactions over rigid processes. This emphasis on collaboration is a universal principle, fostering innovation and efficiency across all industries.

Customer-Centricity: The North Star

At the core of agile principles is a relentless focus on customer satisfaction. In construction, this means ensuring that the final structure aligns perfectly with the client's vision and needs.

In hardware projects, it necessitates constant collaboration with end-users to ensure that systems are tailored to their evolving requirements.

Iterative and Incremental Progress: Agile's Signature Approach

Agile is synonymous with an iterative and incremental approach to work, a practice that is as applicable in construction and hardware as it is in software development.

Construction: Building in Phases

Consider the construction of a hospital. Each phase of the project, from laying the foundation to installing the medical equipment, represents an incremental step toward creating a fully functional healthcare facility. This phased approach embodies the agile principle of iterative progress, ensuring that value is delivered continuously throughout the project's lifecycle.

Hardware Projects: Sprinting to Success

In the world of hardware and network installation, adopting an agile approach means breaking down the project into sprints or iterations. Each sprint delivers a functional component of the network, aligning perfectly with agile's emphasis on delivering working products incrementally.

Agile Values in Practice: A Guiding Light

The values outlined in the Agile Manifesto serve as a guiding light, illuminating the path to success in any domain. They emphasize the importance of individuals and interactions, working products, customer collaboration, and responsiveness to change.

Construction: Collaborating for Innovation

In construction, prioritizing individuals and interactions over processes leads to enhanced collaboration among architects, engineers, and builders. This collaborative spirit fosters innovation and efficiency, resulting in more successful construction projects.

Hardware Projects: Aligning with End-User Needs

In hardware development, the emphasis on customer collaboration ensures that solutions are perfectly aligned with end-user needs. This often involves establishing continuous feedback loops, allowing for constant adaptation to evolving technology requirements.

Lean Thinking

Lean thinking is a management philosophy derived from the Toyota Production System (TPS) that focuses on creating value for the end customer while minimizing waste and optimizing processes. The main objective of lean thinking is to identify and eliminate non-value-adding activities, ensuring that every step in a process adds value to the customer.

Key Principles of Lean Thinking:

1. **Define Value**: Understand what value means to the customer. What are the products or services they are willing to pay for? Understanding customer value is the starting point for lean thinking.
2. **Map the Value Stream**: Analyze the entire process from start to finish, identifying every step involved in the creation and delivery of the product or service. The aim is to map out all the activities that contribute to the value stream and identify any steps that do not add value.

3. **Create Flow**: Once the value stream is identified, the next step is to ensure that value-creating steps flow smoothly without interruptions, delays, or bottlenecks. This could involve reorganizing workstations, improving processes, or removing unnecessary steps.
4. **Establish Pull**: Instead of pushing work through the system based on forecasts or schedules, lean thinking encourages a pull approach, where work is pulled through the system based on customer demand. This helps to reduce overproduction and inventory levels.
5. **Pursue Perfection**: Lean thinking is an ongoing journey towards perfection, where organizations continuously seek ways to reduce waste and improve processes. This involves regularly reviewing and refining processes, encouraging employee feedback, and fostering a culture of continuous improvement.

Benefits of Lean Thinking:

- **Reduced Waste**: By identifying and eliminating non-value-adding activities, organizations can significantly reduce waste in terms of time, resources, and materials.
- **Increased Efficiency**: Processes are streamlined, and flow is improved, leading to faster production times and increased efficiency.
- **Improved Quality**: With a focus on continuous improvement and employee involvement, the quality of products and services is enhanced.
- **Customer Satisfaction**: By delivering value and reducing waste, organizations can offer better products and services at competitive prices, leading to increased customer satisfaction.
- **Employee Engagement**: Employees are encouraged to contribute their ideas and are involved in problem-solving, leading to higher job satisfaction and engagement.

Lean thinking requires a shift in mindset from traditional management practices, emphasizing the importance of customer value, employee involvement, and continuous improvement. It is applicable across various industries and sectors, not just in manufacturing, where it originated.

Definition of Lean Construction

Lean Construction is a production management-based approach to project delivery that aims to maximize value and minimize waste in the construction process. It draws inspiration from the Lean manufacturing principles developed in the automotive industry, focusing on continuous improvement, efficiency, and the elimination of processes that do not add value to the end product.

In recent years, the construction industry has witnessed a significant transformation towards embracing lean practices. Lean construction is centered around minimizing waste and maximizing value, aiming to execute projects with greater efficiency and efficacy.

Lean construction principles are versatile, applicable across various stages of a construction project, spanning from the initial planning and design phases to the construction, commissioning, and maintenance stages. The adoption of lean practices empowers professionals in the construction field to enhance productivity, cut costs, expedite project delivery, and boost customer satisfaction.

A pivotal aspect of lean construction is the integration of diverse stakeholders in the project. This includes professionals such as engineers, architects, project managers, scrum masters, product owners, and more, all of whom collaborate closely to ensure seamless information flow and communication. This collaborative ethos not only fosters innovation but also mitigates misunderstandings and enhances problem-solving throughout the project's lifecycle.

Agile project management stands out as another crucial facet of lean construction practices. Construction professionals can leverage agile approaches to deconstruct projects into smaller, more manageable tasks, facilitating iterative development and ongoing improvement. This strategy ensures that project teams remain agile, capable of swiftly adapting to changes, mitigating risks, and delivering projects on schedule and within budget.

Moreover, lean construction practices underscore the significance of adept supply chain management within the construction sector. Employing agile supply chain management strategies enables professionals to streamline processes, optimize the procurement of materials and equipment, reduce lead times, and diminish waste. This results in smoother project operations, characterized by minimal disruptions and delays.

Beyond construction projects, lean practices are also applicable to hardware development and engineering consulting within the construction industry. Agile engineering processes and principles can be applied to refine the design and development of hardware components, ensuring efficient manufacturing and installation. Additionally, agile principles can enhance project delivery in linear infrastructure development projects, such as roads, railways, and pipelines.

The 7 Wastes in Lean Construction Pertaining to Hardware Type Projects

Transport – Lean Waste

What is it?

Transport waste refers to the unnecessary movement of materials, products, or information that does not add value to the product or service.

Why is it a Problem?

Transport waste can lead to increased costs, longer lead times, and a higher risk of damage to materials. It also contributes to inefficiency, as resources are spent on moving items rather than on value-added activities.

Example in Hardware Projects

Transporting components between different manufacturing facilities or warehouses before they are needed in the assembly process.

Inventory – Lean Waste

What is it?

Inventory waste occurs when there is an excess of materials, work-in-progress, or finished goods that are not immediately required.

Why is it a Problem?

Excess inventory ties up capital, takes up space, and can lead to obsolescence or spoilage. It also hides problems in the production process, such as imbalances in workloads or inefficiencies.

Example in Hardware Projects

Keeping a large stock of components "just in case" they are needed, rather than employing a just-in-time procurement strategy.

Motion – Lean Waste

What is it?

Motion waste refers to unnecessary movements by people or equipment that do not add value to the product or service.

Why is it a Problem?

Motion waste leads to inefficiency, longer production times, and increased wear and tear on equipment. It can also contribute to worker fatigue and increase the risk of injury.

Example in Hardware Projects

Workers having to walk long distances to retrieve tools or materials due to poor workshop layout.

Waiting – Lean Waste

What is it?

Waiting waste occurs when workers or machines are left idle, waiting for materials, instructions, or the completion of a preceding task.

Why is it a Problem?

Waiting leads to lost productivity, longer lead times, and can create a bottleneck in the production process.

Example in Hardware Projects

Delays in the production process due to a lack of synchronization between different stages of production or delayed deliveries of essential components.

Over-Processing – Lean Waste

What is it?

Over-processing waste occurs when more work is done on a product or service than what is required by the customer.

Why is it a Problem?

Over-processing leads to increased costs, longer production times, and can reduce the quality of the final product.

Example in Hardware Projects

Using a high-precision CNC machine for a task that could have been completed with simpler tools.

Over-Production – Lean Waste

What is it?

Over-production waste occurs when more products or services are produced than are required.

Why is it a Problem?

Over-production can lead to excess inventory, increased storage costs, and potential obsolescence. It also hides inefficiencies in the production process.

Example in Hardware Projects

Manufacturing too many parts before they are needed in the assembly process.

Defects – Lean Waste

What is it?

Defects waste occurs when products or services do not meet quality standards and need to be reworked or discarded.

Why is it a Problem?

Defects lead to increased costs, longer production times, and can damage the reputation of the business.

Example in Hardware Projects

Poor workmanship, inadequate quality control, or the use of substandard materials leading to defective components.

Skills – Lean Waste

What is it?

Skills waste occurs when the skills and talents of employees are not fully utilized or are misallocated.

Why is it a Problem?

Underutilizing skills leads to inefficiency, lower employee morale, and can result in a loss of competitive advantage.

Example in Hardware Projects

Assigning a highly skilled technician to perform routine tasks that do not require their level of expertise.

Conclusion: Principles That Transcend Boundaries

In conclusion, the principles of agile are not confined to the digital realm of software; they are versatile, powerful tools that illuminate the path to success in construction, hardware, and a myriad of other industries. By embracing adaptability, fostering collaboration, prioritizing customer-centricity, and adhering to an iterative approach, professionals across diverse fields can unlock the full potential of agility, delivering unparalleled value and excellence. The subsequent chapters will venture deeper into the practical application of these principles, showcasing how agile thinking can transform industries, drive innovation, and redefine what is possible.

Agile Estimation and Planning Techniques

In the dynamic and unpredictable realm of construction, professionals such as engineers, architects, project managers, and others are perpetually challenged to estimate and plan their projects effectively. Traditional methodologies often prove inadequate when confronted with uncertainties, evolving requirements, and stringent deadlines. This is precisely where Agile estimation and planning techniques become invaluable.

Agile practices have witnessed a surge in popularity across construction, hardware, engineering, and linear environments, attributed to their resilience, adaptability, and capacity to deliver value in a flexible, iterative manner.

A crucial component of Agile estimation and planning is the adoption of relative sizing techniques, such as Planning Poker and T-Shirt Sizing. These methods necessitate the collective involvement of the project team, including roles like the scrum master, product owner, and various stakeholders, to allocate relative sizes or effort points to different project tasks or user stories. This collaborative approach fosters more precise effort estimations and aids in task prioritization based on their relative significance and complexity.

Another pivotal technique is Agile release planning. This strategy empowers professionals to schedule and prioritize project deliverables in small, incremental releases, ensuring consistent value delivery to the client throughout the project's lifecycle. By segmenting the project into smaller, more manageable portions, teams can remain agile, adjusting to evolving requirements and making necessary modifications on the fly.

Applying Agile principles such as collaboration, transparency, and continuous improvement can optimize supply chain operations, minimize waste, and bolster overall project efficiency. In addition to these applications, Agile principles are also applicable to engineering consulting, project execution in linear infrastructure development, and hardware development. By adopting Agile approaches, professionals operating in these specialized areas can enhance collaboration, streamline communication, and achieve high-quality outcomes within the defined constraints of time, cost, and scope.

Chapter 2: Universality of Agile Principles: Bridging Industries

1. **Agile in Different Industries**: Provide examples of how Agile principles can be applied in industries outside of software development, such as manufacturing, healthcare, or education.

    - *Guidelines*: Highlight the universality of Agile principles and how they can enhance flexibility, efficiency, and customer satisfaction in various domains.

2. **Challenges in Adopting Agile Across Industries**: Discuss potential challenges and resistance that might be faced when introducing Agile practices in a traditionally non-Agile industry.

    - *Guidelines*: Talk about cultural resistance, lack of understanding, and the need for a mindset shift. Discuss strategies to overcome these challenges.

3. **Benefits Across the Board**: How can adopting Agile principles benefit not just the product development teams, but also other departments like HR, marketing, and sales?

Guidelines: Discuss the holistic benefits of Agile, emphasizing improved communication, faster response to change, and a stronger focus on customer needs.

Chapter 3: Agile Manifesto Values: A Universal Compass for Success

> *"Agile is all about teams working together to produce great software. As an Agile coach, you can help your team go from first steps to running with Agile to unleashing their full Agile potential."*
>
> *– Rachel Davies and Liz Sedley Agile trainers and authors*

Introduction

The Agile Manifesto, a seminal document in the world of agile thinking, outlines a set of values that serve as a compass for guiding projects to success. These values are not just applicable to software development; they transcend industry boundaries, proving their worth in construction, hardware, and other seemingly linear fields. In this chapter, we will embark on a comprehensive exploration of each of these values, illustrating their universal applicability and their power to create successful, customer-centric outcomes.

The Essence of Agile Values

The Agile Manifesto is built upon four core values: Individuals and Interactions over Processes and Tools; Working Product over Comprehensive Documentation; Customer Collaboration over Contract Negotiation; and Responding to Change over Following a Plan. These values encapsulate the essence of agile thinking, emphasizing the importance of human elements, practical results, customer engagement, and adaptability.

Individuals and Interactions over Processes and Tools

At the heart of agile thinking is the belief that the most valuable resource in any project is the people involved. This value underscores the importance of fostering strong, collaborative relationships among team members, clients, and stakeholders.

Construction: Building Together

In the construction industry, the synergy between architects, engineers, and construction workers is paramount. Prioritizing interactions over rigid adherence to processes and tools can lead to innovative solutions, streamlined workflows, and a more responsive approach to project challenges.

Case Study: The Agile Skyscraper

Consider the construction of a skyscraper. The complexity of such a project requires seamless communication and collaboration among all parties involved. By prioritizing individuals and their interactions, the project team can quickly adapt to changes, solve problems collaboratively, and ensure that the building meets the highest standards of quality and safety.

Hardware Projects: Collaborative Innovation

In hardware projects, the rapid pace of technological change demands a flexible, collaborative approach. Emphasizing the importance of individuals and their interactions fosters an environment where technical expertise and innovative solutions can flourish.

Case Study: Agile Network Installation

Imagine a large-scale network installation project. The success of the project hinges on the ability of network engineers, technicians, and stakeholders to work together effectively. By valuing individuals and interactions over rigid processes and tools, the team can adapt to unforeseen challenges, optimize workflows, and deliver a robust, reliable network infrastructure.

Working Product over Comprehensive Documentation

The second Agile Manifesto value advocates for placing higher value on the delivery of a functional, working product over comprehensive documentation. This is not discrediting documentation or its importance. Afterall, on some projects, the documentation is the main deliverable, so the question should be, "how valuable is what we are working on, and dedicating our resources to right now?"

Construction: Tangible Progress

In construction, this value translates to a focus on completing functional phases of the building, ensuring that tangible progress is made toward the final structure. This approach not only provides immediate value to the client but also allows for adjustments and improvements to be made in real-time, based on actual project conditions.

Case Study: Agile Hospital Construction

Consider the construction of a hospital. By adopting an agile approach, the project team can focus on completing functional sections of the hospital, such as emergency rooms or patient wards, before the entire building is finished. This ensures that the community can start benefiting from the new facilities sooner, while also providing valuable feedback that can be used to improve the remaining phases of construction.

Hardware Projects: Early Value Delivery

In hardware projects, delivering a working system, even if not fully complete, can provide immediate value to end-users. This approach aligns with the agile value of prioritizing working products over extensive documentation, ensuring that users can start benefiting from the system as soon as possible.

Case Study: Agile Network Upgrade

Imagine a large-scale network upgrade project. By adopting an agile approach, the project team can focus on delivering functional components of the network incrementally, allowing end-users to start experiencing the benefits of the upgrade before the entire project is complete. This approach not only provides immediate value but also allows for continuous feedback and improvements, ensuring that the final system meets all user requirements.

Customer Collaboration over Contract Negotiation

The third Agile Manifesto value emphasizes the importance of establishing collaborative relationships with customers, moving away from rigid contract negotiations and toward a more flexible, responsive approach to project management.

Construction: Building Client Relationships

In the construction industry, this value translates to a focus on building strong, collaborative relationships with clients. By working closely with clients throughout the project, the construction team can ensure that the final building meets the client's needs, even if those needs change during the project.

Hypothetical Case Study: Agile Commercial Development

Consider the development of a commercial building. By adopting an agile approach, the project team can establish a collaborative relationship with the client, ensuring that their needs and preferences are taken into account throughout the construction process. This approach not only leads to a more successful project but also builds trust and strengthens the client-contractor relationship.

Hardware Projects: Aligning with User Needs

In hardware projects, customer collaboration is crucial for understanding and aligning with evolving technology requirements. By establishing continuous feedback loops and working closely with end-users, the project team can ensure that the hardware solutions meet user needs and expectations.

Hypothetical Case Study: Agile Hardware Implementation

Imagine a complex hardware implementation project. By adopting an agile approach, the project team can collaborate closely with end-users, gathering feedback and making adjustments in real-time to ensure that the hardware system meets all user requirements. This approach not only leads to a more successful project but also builds trust and ensures that the hardware solution is aligned with user needs.

Responding to Change over Following a Plan

The fourth Agile Manifesto value encourages teams to embrace change and remain flexible, rather than rigidly adhering to a predefined plan. This value is crucial for adapting to dynamic project environments and ensuring that the final product meets the highest standards of quality and relevance.

Construction: Adapting to the Unexpected

In the construction industry, unforeseen challenges such as site conditions, material availability, or weather can disrupt plans and timelines. By adopting an agile approach, the construction team can remain flexible, adapting to changes and finding innovative solutions to overcome challenges.

Hypothetical Case Study: Agile Infrastructure Project

Consider a large-scale infrastructure project. By adopting an agile approach, the project team can remain responsive to unforeseen challenges, adapting plans and workflows as needed to ensure that the project stays on track and meets all quality and safety standards. This is what agility is all about. Embedding agile thinking and practices into the process is key for successfully maximizing agile in linear environments.

Hardware Projects: Keeping Pace with Technology

In the world of hardware, the rapid pace of technological change demands a flexible, adaptive approach. By adopting an agile mindset, the project team can remain responsive to changing requirements and emerging technologies, ensuring that the hardware solution stays relevant and meets all user needs.

Hypothetical Case Study: Agile Technology Upgrade

Imagine a technology upgrade project in a large organization. By adopting an agile approach, the project team can remain flexible and responsive, adapting to new technologies and user requirements as they emerge. This approach not only ensures that the technology solution stays relevant but also builds trust and ensures that the organization remains competitive in a rapidly changing technological landscape.

Conclusion: A Guiding Light Across Industries

In conclusion, the Agile Manifesto values serve as a universal compass, guiding professionals in construction, hardware, and a myriad of other industries toward success. By prioritizing individuals and interactions, delivering working products, collaborating with customers, and remaining responsive to change, agile thinking becomes a powerful force for innovation, efficiency, and customer satisfaction. In the upcoming chapters, we will venture deeper into the practical applications of these values, exploring how they can transform projects and industries, irrespective of their traditional rigidity or linearity.

Agile in Business Case Study #2
Title: Agile Transformation in Gillette R&D: A Success Story
This case study provides valuable insights into how a traditional company in the fast-moving consumer goods sector embraced Agile practices to innovate and stay competitive.
This case study ventures into the transformative journey of Gillette's Global R&D division, as it transitioned from traditional waterfall methodologies to Agile Scrum practices for the development of a new razor product. Under the leadership of David Ingram, VP of Global R&D, the team navigated challenges, restructured their approach, and embraced Agile principles to deliver a superior product within an ambitious timeframe. This narrative provides a comprehensive overview of the strategies employed, challenges encountered, and the significant benefits realized through this Agile transformation.
1. Introduction
1.1 **Background**: Gillette, a leader in the razor industry, has traditionally followed waterfall methodologies for product development, launching new products approximately every five to seven years.
1.2 **Objective**: Facing the need for rapid innovation and a shift in consumer preferences towards facial hair, the company aimed to develop a superior shaving product in a significantly reduced timeframe.
2. The Agile Transformation Journey
2.1 **Decision to Adopt Scrum**: The realization that a traditional approach would not suffice led to the decision to adopt Scrum, an Agile framework, known for its iterative and collaborative nature.
2.2 **Engagement with Agile Experts**: Recognizing the need for external expertise, Gillette engaged with Agile Sparks and Agile coach Yuval Yeret to guide the transformation process.
3. Implementation of Scrum
3.1 **Team Reorganization**: The large, multifunctional team of 55 members was restructured into smaller, cross-functional sub-teams, aligning with Scrum guidelines.
3.2 **Alignment on Objectives**: The teams worked towards a unified backlog, ensuring alignment on key objectives and outcomes, and fostering a focus on value delivery and customer satisfaction.
4. Outcomes and Benefits
4.1 **Parallel Creation and Early Tension Identification**: The simultaneous development of various product elements helped in identifying and addressing tensions early in the process.
4.2 **Enhanced Stakeholder Engagement**: Regular interactions with stakeholders, including the CEO, facilitated quicker decision-making and resolution of tensions.
4.3 **Improved Execution Quality and Speed**: The iterative nature of Scrum, combined with regular stakeholder engagement, led to higher quality execution and faster development timelines.
5. Challenges and Organizational Impact
5.1 **Navigating Existing Processes**: Challenges arose when interfacing the Scrum framework with existing waterfall processes, highlighting the need for broader organizational agility.
5.2 **Initiating Broader Transformation**: The success of this project has set the stage for a more extensive Agile transformation at Gillette, aiming to enhance agility and efficiency across the organization.
6. Conclusion
This case study illustrates the successful application of Agile Scrum practices in a traditionally structured organization, resulting in enhanced innovation, efficiency, and product development speed. Gillette's journey serves as a valuable example for other organizations seeking to navigate the complexities of Agile transformation and underscores the potential of Agile philosophies, frameworks and practices in driving organizational success.

Chapter 3: Agile Manifesto Values: A Universal Compass for Success

1. **Deep Dive into Agile Values**: Choose one of the four values of the Agile Manifesto and discuss its relevance and application in today's fast-paced business environment.
   - *Guidelines*: Encourage a deep understanding of the chosen value, providing real-world examples and discussing how it contributes to the success of Agile projects.
2. **Balancing Agile Values**: Discuss a scenario where two Agile values seem to be in conflict, and how a team might navigate this situation.
   - *Guidelines*: Highlight the importance of balance and context in applying Agile values. Encourage critical thinking and problem-solving skills.
3. **Agile Values in Leadership**: How can leaders and managers embody Agile values in their leadership style, and what impact can this have on their teams?
   - *Guidelines*: Discuss the role of leadership in fostering an Agile culture, emphasizing transparency, empowerment, and a focus on delivering value.

Chapter 4: Self-Organization and Inspection: Nurturing Agile Excellence

"

"Although self-organizing is a good term, it has, unfortunately, become confused with anarchy."

– Jim Highsmith (Agile Author)

Introduction

Agile thinking is not just a set of practices; it is a mindset that encourages continuous improvement, innovation, and a commitment to excellence. Central to this mindset are the principles of self-organization and inspection. In this chapter, we will explore these crucial aspects of agile thinking, highlighting their universal relevance and their role in empowering professionals across various industries, including construction, hardware, and other fields that may appear linear and structured at first glance.

Roles and Responsibilities

In the dynamic and complex field of hardware, construction, and engineering, adopting Agile approaches has become increasingly popular as organizations strive to enhance efficiency, improve collaboration, and deliver higher value to their clients. Agile practices, originally rooted in software development, have shown their versatility and effectiveness when adapted to the unique challenges and workflows of these industries. The integration of Agile principles necessitates a clear understanding and redefinition of roles and responsibilities to ensure alignment with Agile values and to facilitate a smooth transition.

The environment in hardware, construction, and engineering projects is typically characterized by long project durations, high capital investments, and a need for precise coordination among diverse professional disciplines. In such a setting, the traditional waterfall project management approach often proves to be rigid and inefficient, leading to delays, cost overruns, and suboptimal outcomes. Agile approaches offer a more flexible, iterative, and collaborative approach, emphasizing continuous improvement, customer satisfaction, and the ability to adapt to changes swiftly.

To successfully implement Agile in this context, it is crucial to define and understand the roles and responsibilities of various professionals involved in the projects. From engineers and architects to project managers and senior leaders, each individual plays a vital role in driving the project forward, ensuring quality, and fostering a culture of collaboration and innovation. The introduction of specific Agile roles such as Scrum Masters and Product Owners also brings a new dimension to the team dynamics, emphasizing the importance of Agile ceremonies, backlog prioritization, and iterative progress.

In the following sections, we will venture deeper into the roles and responsibilities of professionals, engineers, construction workers, architects, project managers, Scrum Masters, Product Owners, and senior managers in a hardware, construction, and engineering Agile environment. This exploration will provide insights into how each role contributes to the Agile ecosystem, ensuring that projects are executed efficiently, obstacles are swiftly addressed, and the final deliverables meet or exceed customer expectations.

1. Professionals:

- **Role**: Professionals in a hardware, construction, and engineering Agile environment could include specialists such as surveyors, planners, and designers. They play a crucial role in providing expert advice and ensuring that the technical aspects of the project meet the required standards.
- **Responsibilities**:
    - Providing technical expertise and guidance.
    - Ensuring compliance with industry standards and regulations.
    - Collaborating with other team members to solve problems and improve processes.
    - Continuously updating their skills and knowledge to stay current in their field.

2. Engineers:

- **Role**: Engineers are responsible for the design, development, and implementation of hardware and construction projects.
- **Responsibilities**:
    - Designing and testing hardware components or construction structures.
    - Collaborating with architects, project managers, and other engineers to ensure project requirements are met.
    - Applying Agile principles to iterate on designs and solutions quickly.
    - Participating in Agile ceremonies and contributing to continuous improvement.

3. Construction Workers:

- **Role**: Construction workers are on the ground, bringing the project plans to life.
- **Responsibilities**:

- Following safety protocols and ensuring a safe working environment.
- Executing construction tasks efficiently and accurately.
- Providing feedback on work processes and potential improvements.
- Adapting to changes in the project plan as dictated by Agile approaches.

4. Architects:

- **Role**: Architects are responsible for the overall design and vision of the hardware or construction project.
- **Responsibilities**:
    - Creating comprehensive design plans that meet client needs and comply with regulations.
    - Collaborating with engineers, construction workers, and project managers to ensure the design is feasible.
    - Adapting designs based on feedback and changes in project scope.
    - Ensuring that the project adheres to Agile principles, promoting flexibility and customer collaboration.

5. Project Managers:

- **Role**: Project managers oversee the project from conception to completion, ensuring that it stays on schedule and within budget.
- **Responsibilities**:
    - Creating and maintaining project schedules and budgets.
    - Communicating with stakeholders and managing their expectations.
    - Facilitating collaboration between different team members and departments.
    - Applying Agile approaches to enhance project flexibility and efficiency.

6. Scrum Masters:

- **Role**: Scrum Masters facilitate Agile practices within the team, ensuring that the Agile process is followed and helping the team to work more effectively.
- **Responsibilities**:
    - Coaching the team on Agile principles and practices.

- Facilitating Agile ceremonies such as daily stand-ups, sprint reviews, and retrospectives.
- Removing impediments that are obstructing the team's progress.
- Working with product owners to ensure the product backlog is well-maintained and prioritized.

7. Product Owners:

- **Role**: Product owners are responsible for defining the vision of the product and ensuring that the team is working on the most valuable features.
- **Responsibilities**:
    - Creating and prioritizing the product backlog.
    - Defining user stories and acceptance criteria.
    - Collaborating with stakeholders to understand their needs and requirements.
    - Ensuring that the team is delivering value to the customers.

8. Senior Managers:

- **Role**: Senior managers provide strategic direction and ensure that the organization's goals are aligned with the Agile transformation.
- **Responsibilities**:
    - Setting the vision and strategy for Agile adoption within the organization.
    - Providing resources and support for Agile teams.
    - Ensuring that the organization's culture is conducive to Agile practices.
    - Monitoring the progress of Agile teams and helping to resolve any issues that arise.

In a hardware, construction, and engineering Agile environment, these roles and responsibilities are crucial for ensuring that projects are completed efficiently, safely, and to the highest quality standards. Collaboration, continuous improvement, and customer focus are key elements that all these roles contribute to, fostering a successful Agile transformation.

Self-Organization: Unleashing Team Potential

Self-organization is a fundamental pillar of agile thinking, advocating for teams that are empowered to manage their own work, make decisions collaboratively, and take ownership of their projects.

Construction: Building from Within

In the construction industry, the concept of self-organization challenges traditional hierarchical structures, encouraging teams on the ground to take initiative, solve problems proactively, and contribute to the project's success.

Empowering the Workforce

When agile-minded construction workers, engineers, and architects are empowered to make decisions and manage their work, the entire project benefits. High-synergy teams find more efficient ways to build, address challenges as they arise, and ensure that quality standards are met consistently.

Hardware Projects: Adapting with Agility

In the realm of hardware, self-organization is equally critical. Teams that are empowered to adapt to changing requirements, emerging technologies, and customer needs can respond with agility, ensuring that the final product is both high-quality and fit-for-purpose.

Responsive and Resilient

Self-organizing hardware teams are better equipped to navigate the fast-paced world of technology, responding to changes and challenges with resilience and innovation.

Daily Inspection: Keeping Agile on Track

Frequent inspection is a hallmark of agile thinking, ensuring that teams are aligned, issues are addressed promptly, and the project stays on course.

Construction: Aligning and Adapting

In construction, daily inspection meetings serve as a platform for team alignment, issue resolution, and continuous adaptation.

Building a Rhythm of Excellence

By instituting a rhythm of daily inspections, construction teams ensure that everyone is on the same page, potential issues are identified early, and necessary adjustments are made in a timely manner.

Hardware Projects: Ensuring Quality and Progress

In hardware installations, daily inspections play a crucial role in maintaining project momentum, ensuring quality, and addressing issues as they arise.

A Proactive Approach

Through daily inspections, hardware teams can proactively identify potential issues, ensuring that the project stays on track and meets the highest standards of quality and reliability.

Iterative Inspection: Refining Agile Practices

Agile thinking encourages inspection not just at the daily level, but also iteratively, at various points throughout the project's lifecycle.

Construction: Phase-by-Phase Excellence

In construction, iterative inspections at different project phases ensure that each stage of the build meets quality standards before moving on to the next.

Foundation to Finish

From foundation inspections to final walkthroughs, iterative inspections in construction ensure that every aspect of the build is up to standard, resulting in a final structure that is safe, sound, and successful.

Hardware Projects: Ensuring System Reliability

In network installations and hardware projects, iterative inspections help identify potential issues early, allowing for timely adjustments and ensuring the reliability and security of the system.

A Systematic Approach to Quality

By conducting inspections at various stages of the hardware project, teams can ensure that each component meets quality standards, resulting in a robust, reliable final product.

The Agile Mindset and Inspection

At its core, agile thinking is about continuous improvement, driven by regular inspection and a willingness to adapt.

Construction: Innovating for Success

In construction, a commitment to continuous inspection and adaptation can lead to innovative solutions, cost savings, reduced waste, and enhanced safety.

Building Better

By embracing the agile mindset, construction teams can transform the way they work, driving innovation and excellence from the ground up.

Hardware Projects: Staying Ahead of the Curve

In hardware installations, frequent inspections help teams stay ahead of technological advancements, respond proactively to security threats, and ensure that the system remains up-to-date and secure.

A Proactive Approach to Technology

By adopting an agile mindset, hardware teams can ensure that they are always ready to respond to the latest technological challenges, keeping their systems secure, reliable, and cutting-edge.

Conclusion: Cultivating Agile Mastery

In conclusion, self-organization and inspection are vital components of agile excellence, applicable across industries and essential for driving innovation, efficiency, and quality. Teams that embrace these principles are better equipped to adapt to change, find innovative solutions, and continuously improve their work. As we move forward, we will explore real-world applications of these agile principles in construction, hardware, and other industries, demonstrating how they empower professionals to achieve excellence and embrace agile thinking in all its forms.

Chapter 4: Self-Organization and Inspection: Nurturing Agile Excellence

1. **Benefits and Challenges of Self-Organization**: List the benefits of having a self-organizing team, and discuss potential challenges that might arise.
   - *Guidelines*: Emphasize the empowerment, innovation, and accountability that come with self-organization, while also addressing potential issues like lack of direction or conflicts.
2. **Role of Inspection in Agile**: How does regular inspection contribute to Agile excellence, and what are some effective ways to implement it in a team?
   - *Guidelines*: Discuss the importance of transparency, continuous improvement, and adapting to change. Provide examples of inspection practices like reviews and retrospectives.
3. **Balancing Autonomy and Guidance**: How can leaders strike the right balance between giving their teams autonomy and providing necessary guidance and support?
   - *Guidelines*: Talk about the importance of trust, clear communication, and setting the right expectations. Discuss how leaders can be enablers rather than controllers.

Chapter 5: Agile Beyond Frameworks: Embracing the Agile Mindset

> *"Inside every large program, there is a small program trying to get out."*
>
> *– C.A.R. Hoare (Computer scientist)*

Introduction

Agile is a term that has transcended its origins in software development, becoming a buzzword in industries ranging from construction to hardware. However, there is a common misconception that being agile is synonymous with adopting specific frameworks or methodologies. In reality, the true essence of agility lies in the mindset. In this chapter, we will unravel the concept of the agile mindset, illustrating how it can be applied universally across various fields, transforming traditional practices and fostering a culture of collaboration, adaptability, and continuous improvement.

The Agile Mindset: A Paradigm Shift

Agility represents a fundamental shift in how we approach work, challenges, and collaboration. It is a mindset that values adaptability, openness to change, and a relentless focus on delivering value.

Construction: Building Collaboratively

In the construction industry, an agile mindset transforms the traditional hierarchical structure, fostering a culture where architects, builders, and contractors collaborate seamlessly.

"Innovation distinguishes between a leader and a follower." - Steve Jobs

Embracing an agile mindset in construction means leading with innovation, finding new ways to solve old problems, and optimizing processes for efficiency and effectiveness.

Hardware Projects: Adapting with Precision

In the realm of hardware, agility is synonymous with adaptability. An agile mindset ensures that systems remain relevant, valuable, and aligned with the ever-evolving landscape of technology and customer needs.

"It's fine to celebrate success, but it is more important to heed the lessons of failure."

- Bill Gates

Agile thinking in hardware projects involves learning from past experiences, staying attuned to technological advancements, and being ready to pivot when necessary.

The Agile Mindset vs. Frameworks

Agility is not confined to the boundaries of frameworks or methodologies; it is a way of thinking and behaving that transcends these external constructs.

Construction: Agility Without Borders

In construction, an agile mindset can thrive even in the absence of formal agile frameworks. It's about being responsive to changing client requirements, site conditions, and finding ways to optimize processes on the fly.

"Do not wait; the time will never be 'just right.' Start where you stand, and work with whatever tools you may have at your command, and better tools will be found as you go along."

- George S. Patton

This quote encapsulates the essence of agility in construction – starting with a collaborative mindset and continuously improving processes and interactions.

Hardware Projects: Beyond Methodologies

In hardware installations, agility is about more than just following a set methodology. It's about fostering a culture of collaboration, being open to technological changes, and responding promptly to evolving needs.

"The only way to do great work is to love what you do."

- Steve Jobs

Loving what you do in the hardware world means staying passionate about technology, being curious about new advancements, and being ready to adapt to the changing tech landscape.

Agile Values in Action

The values outlined in the Agile Manifesto serve as a universal guide, applicable across industries and independent of specific agile frameworks.

Construction: Teamwork and Innovation

Emphasizing individuals and interactions over processes and tools leads to a more collaborative and innovative construction environment.

"Your work is going to fill a large part of your life, and the only way to be truly satisfied is to do what you believe is great work."

- Steve Jobs

In construction, doing great work means fostering a culture of teamwork, open communication, and continuous improvement.

Hardware Projects: Customer-Centric Adaptability

In the hardware realm, customer collaboration and responding to change are paramount. An agile mindset ensures that systems are developed with the end-user in mind, and are adaptable to their evolving needs.

"It's not about ideas. It's about making ideas happen."

- Scott Belsky

Making ideas happen in hardware projects requires a relentless focus on customer collaboration, feedback, and adaptability.

The Role of Agile Thinking

Agile thinking is about more than just adopting agile practices; it's a mindset that encourages continuous improvement, innovation, and a commitment to excellence.

Construction: A Culture of Continuous Improvement

An agile thinker in construction is always looking for ways to optimize processes, enhance collaboration, and deliver value to the client.

"Success is not the key to happiness. Happiness is the key to success. If you love what you are doing, you will be successful."

- Albert Schweitzer

Finding happiness in the continuous improvement and collaborative nature of construction leads to successful projects and satisfied clients.

Hardware Projects: Staying Ahead of the Curve

In hardware, agile thinking involves being proactive, staying ahead of technological trends, and ensuring that systems are developed with future needs in mind.

"Innovation is the ability to see change as an opportunity – not a threat."

- Steve Jobs

Seeing change as an opportunity is crucial in the fast-paced world of hardware, ensuring that systems remain cutting-edge and relevant.

Conclusion: Cultivating an Agile Culture

In conclusion, agility is not confined to specific frameworks or methodologies; it is a mindset that can transform industries, from construction to hardware and beyond. Embracing the agile mindset means fostering a culture of collaboration, adaptability, and continuous improvement, leading to innovative solutions and successful outcomes.

"Your work is going to fill a large part of your life, and the only way to be truly satisfied is to do what you believe is great work. And the only way to do great work is to love what you do."

- Steve Jobs

In the chapters to come, we will venture deeper into practical applications of the agile mindset, showcasing real-world examples of how professionals across various fields are embracing agility to excel and innovate.

Chapter 5: Agile Beyond Frameworks: Embracing the Agile Mindset

1. **Agile Mindset in Daily Work**: Provide examples of how an individual can apply the Agile mindset in their daily tasks, regardless of their role or industry.
    - *Guidelines*: Encourage thinking beyond frameworks and methodologies, focusing on the core principles of Agile in everyday work.
2. **Overcoming Framework Limitations**: Discuss how teams can overcome the limitations of a specific Agile framework by embracing a broader Agile mindset.
    - *Guidelines*: Highlight the importance of adaptability, continuous learning, and focusing on the underlying principles rather than getting stuck in a specific framework.
3. **Agile Mindset in Non-Project Work**: How can the Agile mindset be applied in areas of work that are not project-based, such as operations or support functions?
    - *Guidelines*: Discuss the universality of the Agile mindset, providing examples of how its principles can enhance efficiency, responsiveness, and customer satisfaction in various work contexts.

Chapter 6: Network Implementation with Agile: Bridging the Digital Divide

"

"Change is not for its own sake, it's the means to the end, not the end."

– Pearl Zhu (Author)

Introduction

In the digital age, network implementation is a critical component of any large corporate campus. It ensures seamless communication, data transfer, and overall operational efficiency. However, implementing a network across a vast campus is no small feat. It requires meticulous planning, coordination among various teams, and a deep understanding of both the technical and business requirements. But what if we approached this seemingly linear and structured task with an agile mindset? In this chapter, we will venture into the transformative power of agile in the realm of network implementation, showcasing how it can bridge the digital divide and lead to more efficient, adaptable, and user-centric outcomes.

The Challenge of Network Implementation

Implementing a network in a large corporate campus is akin to building the digital backbone of the organization. It involves:

- **Infrastructure Planning:** Determining the layout of the network, including routers, switches, servers, and other hardware components.
- **Connectivity:** Ensuring seamless connectivity across buildings, departments, and even remote locations.
- **Security:** Implementing robust security measures to protect against external threats and internal vulnerabilities.
- **Integration:** Integrating the new network with existing systems and ensuring compatibility with various devices and software applications.

Given the complexity and scale of such projects, they often follow a linear and phased approach. However, this can lead to challenges such as delays, cost overruns, and a lack of flexibility to adapt to changing requirements.

Thorough Planning with Agile Principles

Agile offers a fresh perspective on network implementation. Instead of a rigid, step-by-step approach, agile emphasizes flexibility, collaboration, and continuous feedback. Here's how agile principles can be incorporated into the planning phase:

Defining Scope with Flexibility

While it's essential to have a clear project scope, agile recognizes that requirements can evolve. By adopting an agile mindset, teams can remain open to changes, ensuring that the network meets the organization's current and future needs.

Team Charter for Collaboration

A team charter, in the agile context, is more than just a document outlining roles and responsibilities. It's a commitment to collaboration, open communication, and shared goals. By establishing a team charter, network implementation teams can foster a culture of teamwork and innovation.

Iterative Sprint Approach for Incremental Progress

Agile is all about incremental progress. Instead of waiting for the entire network to be implemented, agile teams can break down the project into smaller sprints or phases, delivering tangible results after each sprint. This approach offers several benefits:

- **Faster Feedback:** By implementing the network in phases, teams can gather feedback from end-users and make necessary adjustments in subsequent sprints.
- **Risk Mitigation:** Smaller sprints allow teams to identify and address potential issues early on, reducing the risk of major setbacks later in the project.
- **Adaptability:** An iterative approach provides the flexibility to adapt to changing requirements, whether it's integrating new technologies or accommodating organizational changes.

The Kanban Advantage

Agile construction practices are revolutionizing project execution, fostering enhanced collaboration, continuous improvement, and adaptability. The Kanban method, in particular, aligns seamlessly with these principles, offering a framework for visualizing work, limiting work in progress (WIP), and optimizing workflow.

In the realm of construction projects, the ability to visualize work is paramount for effective project management. Utilizing Kanban boards, professionals gain the ability to track task progress, pinpoint bottlenecks, and distribute resources efficiently. This level of transparency bolsters collaboration and facilitates swift decision-making, culminating in superior project outcomes.

Addressing WIP is a critical component of the Kanban method, preventing resource overload and ensuring timely task completion. By establishing WIP limits at each stage of the construction process, professionals can uphold a consistent workflow and sidestep potential delays.

Moreover, the Kanban method champions continuous improvement through "Kaizen." Given the complexity and potential for unexpected challenges in construction projects, the ability for teams to perpetually refine and enhance their processes is invaluable. Kanban empowers teams to identify improvement opportunities and incrementally implement changes.

Specific sectors within the construction industry, such as linear infrastructure development, engineering consulting, and hardware manufacturing, stand to gain significantly from adopting the Kanban method. Integrating Agile principles into these specialized areas can streamline project execution, elevate supply chain management, and boost overall efficiency.

Utilizing a framework like a Kanban board fosters transparency, allowing teams to visually track the progress of tasks and workflow, which in turn enhances numerous aspects of project management:

- ✓ **Faster Feedback:** With the clarity provided by a Kanban board, it's easier to implement projects in phases. This phased approach facilitates quicker feedback from end-users as they interact with each iteration. Teams can then utilize this feedback to make necessary adjustments in the upcoming sprints, ensuring that the project remains aligned with user needs and expectations.
- ✓ **Risk Mitigation:** Breaking down the project into smaller sprints, visually represented on a Kanban board, allows teams to identify and address potential issues at an early stage. This proactive identification and resolution of issues significantly reduce the risk of major setbacks as the project progresses. The visual nature of the Kanban board helps in spotting bottlenecks or other hurdles in real-time, allowing for timely intervention.
- ✓ **Adaptability:** An iterative approach, supported by the visual management of a Kanban board, provides the much-needed flexibility to adapt to changing requirements. Whether it's integrating new technologies or accommodating organizational changes, teams can adjust their plans and actions swiftly. The visual cues from the Kanban board help in understanding the impact of changes and making informed decisions on how to navigate them.

The Kanban board acts as a mirror, reflecting the current state of the project in a clear, accessible manner. This transparency is instrumental in fostering a culture of continuous improvement, promoting a shared understanding among team members, and aligning actions with project goals. Through this lens, teams can navigate the complexities and dynamic nature of projects with a clear sight, making informed, timely decisions that drive value and success.

Moreover, the Kanban method actively promotes a culture of continuous improvement through the "Kaizen" philosophy. Construction projects, known for their complexity and susceptibility to unforeseen challenges, can greatly benefit from this approach. Kanban enables teams to perpetually refine and optimize their workflows, ensuring that areas in need of improvement are promptly identified and addressed through incremental changes.

Linear infrastructure development, engineering consulting, and hardware manufacturing represent specialized segments within the construction sector that stand to gain substantially from the Kanban method. The integration of Agile principles within these domains can significantly streamline project execution, bolster supply chain management, and enhance overall operational efficiency.

In summation, the Kanban method holds remarkable potential to transform the landscape of construction projects. Its seamless integration into Agile construction practices, coupled with its applicability in hardware development, engineering, and linear infrastructure projects, paves the way for marked advancements in project delivery, collaborative efforts, and operational efficiency. By adopting the Kanban method, professionals within the construction industry are poised to achieve unprecedented levels of productivity and success.

Regular Team Inspections for Continuous Improvement

One of the hallmarks of agile is the emphasis on continuous improvement. Regular team inspections, such as daily stand-ups and sprint retrospectives, play a crucial role in this:

Daily Scrums / Daily Stand-Ups for Alignment

Daily stand-up meetings are a staple of Agile approaches. They provide an opportunity for team members to share updates, discuss challenges, and align on priorities. In the context of network implementation, daily stand-ups can ensure that the project stays on track and that any issues are addressed promptly.

Sprint Retrospectives for Learning

At the end of each sprint, teams can conduct a retrospective to reflect on what went well, what could be improved, and how to enhance the process in the next sprint. This practice fosters a culture of continuous learning and improvement.

Customer Collaboration for User-Centric Outcomes

In agile, the customer is at the heart of everything. For network implementation, the "customers" are the end-users – the employees, departments, and teams that will use the network. Collaborating with them throughout the project can lead to better outcomes:

- **Feedback Loops:** Regularly engaging with end-users to gather feedback ensures that the network meets their needs and expectations.
- **Transparency:** Keeping stakeholders informed about the project's progress fosters trust and ensures that everyone is on the same page.

Creative Value Delivery: Beyond the Network

While the primary goal of network implementation is to establish a robust and efficient network, agile teams can deliver additional value along the way:

- **Documentation:** Providing comprehensive documentation, including network schematics, configurations, and user guides, can enhance the user experience and simplify future maintenance.
- **Training:** Offering training sessions for end-users can ensure that they make the most of the new network and are aware of best practices for security and efficiency.

Conclusion: Agile as the Bridge to Digital Excellence

In conclusion, agile offers a transformative approach to network implementation in large corporate campuses. By emphasizing flexibility, collaboration, and user-centricity, agile bridges the digital divide, leading to networks that are not only technically robust but also aligned with the organization's evolving needs. As we move forward, it's clear that the principles of agile have a place beyond software development, offering a roadmap to excellence in diverse fields, including the intricate world of network implementation.

Chapter 6: Network Implementation with Agile: Bridging the Digital Divide

1. **Agile in Network Implementation**: Discuss how Agile principles can be applied in the context of network implementation and what unique challenges might arise.
    - *Guidelines*: Focus on the specificities of network implementation, discussing how Agile can enhance flexibility, risk management, and customer collaboration.
2. **Bridging the Digital Divide with Agile**: How can Agile practices contribute to bridging the digital divide, and what role does collaboration play in this context?
    - *Guidelines*: Discuss the social impact of Agile practices, emphasizing the importance of inclusivity, continuous learning, and community engagement.
3. **Agile in Large-Scale Network Projects**: Explore the application of Agile in large-scale network projects, discussing potential benefits and strategies for overcoming challenges.
    - *Guidelines*: Highlight the scalability of Agile, providing examples of how it can be adapted to suit large and complex projects while maintaining its core principles.

Chapter 7: Creative Value Delivery: Unconventional Approaches to Agile Success

"

"A vision should be judged by the clarity of its values, not the clarity of its implementation path."

– Donella H. Meadows (author of Thinking in Systems: A Primer)

Introduction

In the world of construction and hardware, the path to delivering value can often seem linear and predetermined. However, the agile mindset challenges this notion, urging professionals to think outside the box and find creative ways to deliver value at every stage of a project. This chapter ventures into the unconventional approaches that can lead to agile success in industries where the traditional Agile approaches might seem out of place.

Challenges in Value Delivery

Delivering value in construction and hardware projects is inherently challenging. The tangible nature of these industries means that partial products – a half-built building or an incomplete hardware system – often don't offer immediate value to clients or end-users. This poses a unique challenge: How can teams deliver value incrementally in such environments?

"It's not about being perfect. It's about effort. And when you bring that effort every single day, that's where transformation happens. That's how change occurs."

- Jillian Michaels

Think Beyond the Product

The key to agile success in these industries is to broaden the definition of value. It's not just about the final product but the entire journey.

Schematics and Documentation

In construction projects, while a building phase is underway, offering detailed schematics, blueprints, and documentation can provide immense value to stakeholders. It gives them clarity, ensures transparency, and empowers them to make informed decisions.

"Documentation is a love letter that you write to your future self."

- Damian Conway

Progress Updates

For hardware projects, especially those that span over extended periods, regular progress updates can be invaluable. These updates, accompanied by demonstrations or prototypes, offer stakeholders a glimpse into the project's evolution and assure them of the team's capabilities.

"Progress lies not in enhancing what is, but in advancing toward what will be."

- Khalil Gibran

Value in Unexpected Places

Sometimes, value can be found in the most unexpected places. Let's discuss this.

Quality Assurance

In construction, ensuring that each phase of the project meets the highest quality standards can be a significant value driver. Regular inspections, quality checks, and adherence to standards can boost stakeholder confidence and ensure the project's long-term success.

"Quality is never an accident; it is always the result of intelligent effort."

- John Ruskin

Security Measures

In the realm of hardware, with the increasing threats to digital security, implementing robust security measures at every stage can be a game-changer. Even if the system isn't fully operational, ensuring its security can offer peace of mind to stakeholders.

"The only truly secure system is one that is powered off, cast in a block of concrete and sealed in a lead-lined room with armed guards."

- Gene Spafford

Iterate and Adapt

Agility is all about iteration and adaptation. Let's discuss this in more detail.

Continuous Improvement

Encourage teams to adopt a mindset of continuous improvement. Whether it's finding more efficient construction techniques or optimizing hardware configurations, there's always room for enhancement.

"We cannot become what we want by remaining what we are."

- Max Depree

Feedback-Driven Iteration

Soliciting feedback isn't just for software projects. In construction and hardware, understanding the needs, preferences, and concerns of end-users can guide the project's direction, ensuring it aligns with their expectations.

"We all need people who will give us feedback. That's how we improve."

- Bill Gates

Customized Value Delivery

Every project is unique, and so is the value it offers. Let's expand on this.

Client-Specific Value

In construction, understanding the client's vision and requirements can lead to customized value delivery. It could be in the form of design modifications, material choices, or even project timelines.

User-Centric Value

For hardware projects, focusing on the end-user's experience can guide value delivery. Whether it's user-friendly interfaces, compatibility considerations, or performance optimizations, a user-centric approach ensures the project's success.

"You've got to start with the customer experience and work back toward the technology, not the other way around."

- Steve Jobs

Incremental Value Delivery vs. Potentially Shippable Increment Delivery

In the Agile philosophy, particularly when applied to hardware, construction, and engineering, the concepts of incremental value delivery and incremental shippable increment delivery hold distinct meanings and applications. Understanding the nuances between these two approaches is crucial for professionals aiming to optimize project outcomes and client satisfaction.

Potentially Shippable Increment Delivery

Potentially shippable increment delivery is rooted in the principle of breaking down the project into smaller, manageable parts that can be completed, tested, and potentially shipped to the customer. In hardware and construction, this could manifest as completing a specific component of a larger machine or finishing a particular section of a building.

This approach aligns with Agile's emphasis on continuous delivery and feedback, allowing teams to identify issues early, adjust plans swiftly, and ensure that every piece of the project aligns with the overall vision and quality standards. It fosters a sense of progress and momentum, as tangible results are produced regularly.

However, in industries like construction and hardware development, where components may be interdependent or require a certain level of completion before they are functional or valuable, the focus on delivering shippable increments might need to be balanced with a broader view of value delivery.

Incremental Value Delivery

Incremental value delivery shifts the focus from delivering tangible, shippable increments to ensuring that every stage of the project contributes discernible value to the client or end-user. This could be in the form of risk reduction, knowledge gain, client satisfaction, or alignment with long-term goals.

In contexts where a partial product might not hold intrinsic value or functionality—such as a bridge's foundation without the superstructure or a machine's internal mechanism without its casing—the emphasis should be on how the work contributes to the project's overall success and value proposition.

This approach resonates with the Agile mindset, as it encourages teams to continuously reassess and realign their work with the ultimate goal of delivering maximum value, even if the immediate output isn't a standalone, functional product.

Balancing the Approaches

In hardware, construction, and engineering, professionals must navigate the complex landscape of project requirements, client expectations, and the technical realities of their work. The choice between focusing on incremental shippable increments or incremental value delivery should be informed by the nature of the project, the needs of the client, and the stage of development.

1. **Complex and Interdependent Projects**: In scenarios where components are highly interdependent, focusing on incremental value delivery might be more appropriate. Ensuring that each stage of work lays a solid foundation for the next, even if it doesn't result in a shippable product, can lead to a more successful and valuable outcome.
2. **Client-Centric Focus**: Adopting a client-centric focus encourages teams to prioritize activities that bring the most value to the client, even if they don't result in immediate, tangible deliverables. This could involve investing time in design refinement, prototyping, or risk mitigation activities.
3. **Flexibility and Adaptation**: Embracing Agile's core principles of flexibility and adaptation means being open to reassessing the focus of the project as it evolves. Teams should be prepared to shift between incremental shippable increment delivery and incremental value delivery as the context and needs of the project change.

The role of the product owner and the team in an Agile environment, especially within the realms of hardware, construction, and engineering, is pivotal in ensuring that value is not just a buzzword, but a tangible outcome of every iteration and phase of the project. Their responsibilities extend beyond the mere delivery of product increments, delving deep into the realms of value assessment, clarification, expectation setting, and delivery.

Assessing and Analyzing Value

The product owner, with the support of the team, plays a crucial role in meticulously assessing and analyzing what constitutes value for the client and the end-users. This process involves:

1. **Understanding Client Needs**: Engaging in continuous dialogue with the client to comprehend their expectations, goals, and vision for the project.
2. **Market Research**: Conducting thorough market research to understand industry standards, trends, and innovations that could add value to the project.
3. **Risk Assessment**: Identifying potential risks and uncertainties and evaluating their impact on the project's value proposition.

This comprehensive analysis ensures that the team has a holistic understanding of what value means in the context of the specific project, enabling them to make informed decisions and prioritize work accordingly.

Clarifying Value

Once value has been assessed and analyzed, the next step is to clarify and define it in clear, actionable terms. This involves:

1. **Setting Clear Objectives**: Defining clear and concise objectives that align with the identified value parameters.
2. **Prioritizing Work**: Prioritizing tasks and features based on their potential to deliver value, ensuring that the most critical aspects are addressed first.
3. **Creating a Value Roadmap**: Developing a roadmap that outlines how value will be delivered throughout the project, setting clear milestones and checkpoints.

This clarity ensures that the entire team is aligned on what needs to be achieved and why, fostering a shared understanding and commitment to value delivery.

Setting Expectations for Value

Managing expectations is a critical aspect of delivering value. The product owner must:

1. **Communicate Clearly**: Ensure that the client and all stakeholders have a clear understanding of what value will be delivered, and when.
2. **Manage Expectations**: Proactively manage expectations, especially when dealing with complex projects where value delivery might not always be linear or tangible in the early stages.
3. **Build Trust**: Establish and maintain trust with the client and stakeholders through transparency, honesty, and consistent delivery on promises.

Delivering Value

Finally, the onus is on the product owner and the team to ensure that value is not just promised, but delivered. This requires:

1. **Agile Execution**: Implementing Agile practices diligently, ensuring that work is done in iterative, manageable chunks that contribute to the overall value.
2. **Continuous Improvement**: Embracing the principle of Kaizen, or continuous improvement, to constantly refine processes, practices, and outcomes for enhanced value delivery.
3. **Feedback Loops**: Establishing robust feedback loops with the client and stakeholders to ensure that the delivered value aligns with expectations and to make adjustments as necessary.

Agile Ingenuity: Mastering the Art of Innovative Work Breakdown

Breaking down work into smaller, more manageable pieces is a cornerstone of Agile methodology, and it requires a blend of creativity, innovation, and strategic thinking. This process, often referred to as decomposition or work breakdown, is crucial in identifying opportunities for incremental value delivery. It challenges the traditional notion of delivering a project as a single, monolithic entity and instead promotes a more flexible, responsive approach.

The Art of Decomposition

1. **Creative Problem-Solving**: The team must engage in creative problem-solving sessions, brainstorming ways to dissect the project into smaller, more manageable parts. This involves looking at the project from different angles and considering various perspectives to uncover hidden opportunities for incremental delivery.
2. **Value-Driven Focus**: Each piece of the work breakdown should be assessed based on the value it delivers. The team should prioritize tasks that offer immediate and tangible benefits to the client or end-users, ensuring that every increment of work contributes positively to the project's overall value proposition.
3. **Innovative Delivery Strategies**: Adopting innovative delivery strategies can help in identifying novel ways to break down work. For instance, leveraging technology, modular design, or prefabrication techniques in construction and engineering projects can enable more granular delivery of value.

Case Study: CitiGroup at 33 Canada Square, Canary Wharf

Reflecting on my time at CitiGroup at Canary Wharf, where I was Archiving Manager in the early 2000s, the gradual inhabitation of the building serves as a prime example of incremental value delivery. The building was not made functional in one go; instead, services and departments were moved in progressively, ensuring that value was delivered incrementally.

1. **Phased Approach**: Adopting a phased approach to occupancy, where functionality was made available on lower floors while the higher floors were still being worked on, ensured a continuous delivery of value. This approach allowed for immediate utilization of available space, contributing to the overall functionality and efficiency of the building.
2. **Prioritizing Critical Functions**: Departments that were crucial for the building's operation were moved in first, ensuring that the most essential services were up and running. This strategic prioritization ensured that value was delivered where it was needed most, right from the start.
3. **Adapting to Change**: As other floors, such as the trade floor and legal departments, were progressively moved in, the team had to remain adaptable and responsive to changing requirements and circumstances. This agility ensured that the project remained on track, and value was continuously delivered throughout the process.

The art of breaking down work into smaller, value-driven increments is a critical skill in Agile environments, particularly in hardware, construction, and engineering projects. It requires a combination of creativity, innovation, and strategic thinking, with a relentless focus on delivering tangible value at every stage of the project. By adopting this approach, teams can ensure that they are not just delivering products or services, but are actively contributing to the overall success and value proposition of the project.

Conclusion: The Art of Value Delivery

In the end, delivering value is an art. It requires creativity, adaptability, and a deep understanding of stakeholder needs. While construction and hardware projects present their unique challenges, the principles of agile offer a roadmap to success. By thinking beyond the product, finding value in unexpected places, iterating based on feedback, and customizing value delivery, professionals can ensure project success and stakeholder satisfaction.

While incremental shippable increment delivery provides a clear and tangible sense of progress, it is the philosophy of incremental value delivery that often resonates more profoundly in hardware, construction, and engineering projects. By adopting a flexible, client-centric approach, professionals in these fields can ensure that they are not just delivering parts of a product, but are consistently contributing to the overall value and success of the project.

The product owner and the team are the linchpins in the Agile value delivery process, especially in hardware, construction, and engineering projects. Their ability to assess, analyze, clarify, set expectations for, and ultimately deliver value is what transforms Agile from a methodology to a mindset, ensuring that the focus remains steadfastly on delivering tangible, meaningful outcomes that meet and exceed client expectations.

"Innovation distinguishes between a leader and a follower."

\- Steve Jobs

Chapter 7: Creative Value Delivery: Unconventional Approaches to Agile Success

1. **Unconventional Agile Practices**: Provide examples of unconventional or innovative Agile practices, and discuss how they contribute to project success.
    - *Guidelines*: Encourage creativity and open-mindedness, highlighting how thinking outside the box can lead to more effective and efficient Agile practices.
2. **Balancing Creativity and Structure**: How can teams balance the need for creative, unconventional approaches with the need for structure and consistency in Agile practices?
    - *Guidelines*: Discuss the importance of finding the right balance, providing examples of how teams can maintain a structured yet flexible approach to Agile.
3. **Measuring the Impact of Creativity**: Explore ways in which the impact of creative and unconventional Agile practices can be measured and evaluated.
    - *Guidelines*: Emphasize the importance of continuous improvement, discussing how teams can use metrics and feedback to assess the effectiveness of their practices and make informed adjustments.
4. **Innovative Work Breakdown:** Explore ways in which your project work can be broken down move innovatively.
    - *Guidelines:* Emphasize the importance of value, benefits and outcome over output.

Chapter 8: The Customer's Perspective: Value from Their Eyes

> *"There is nothing so useless as doing efficiently that which should " not be done at all."*
>
> *– Peter Drucker (Management consultant and author)*

Introduction

In the realm of project management, understanding the customer's perspective is paramount. Agile approaches, with their inherent flexibility and responsiveness, place the customer at the epicenter of all project activities. This chapter aims to unravel the layers of customer-centric approaches in various industries, with a particular focus on construction and hardware projects. We will venture into the nuances of aligning project goals with customer needs, the indispensable role of continuous feedback, and how these principles of agility transcend industry boundaries.

The Agile Focus on Customer Perspective

Agile approaches are not just a set of practices; they represent a shift in mindset, emphasizing the importance of viewing the project through the customer's lens. This focus is universally applicable and resonates across various domains, proving that agility is not confined to the realms of software development.

Customer-Centricity in Construction

In the construction industry, understanding and aligning with the client's perspective is not just beneficial; it is essential. Engaging in regular, open dialogues with the client ensures that their evolving needs and expectations are understood and considered. It's about creating a synergy between the builders and the clients, ensuring that the final structure not only stands tall but also aligns perfectly with the client's vision.

Moreover, the end-users, the occupants of the constructed space, cannot be overlooked. Their comfort, convenience, and safety are paramount. Designing and constructing with the end-user in mind ensures that the building is not just a structure but a space that resonates with those who inhabit it.

Customer-Centricity in Hardware Projects

Similarly, in hardware projects, the end-user is king. Engaging with them, understanding their interactions with the hardware, and gathering their feedback becomes a crucial part of the development process. It's about ensuring that the hardware not only functions optimally but also aligns with the end-user's expectations and evolving needs.

Staying attuned to technological advancements and trends ensures that the hardware remains relevant and continues to serve the end-user's needs, even as those needs change and evolve.

Value: From the Customer's Eyes

What we, as project managers and developers, perceive as valuable might not always align with what the customer values. This potential disconnect highlights the critical importance of a customer-centric approach.

In Construction Projects

In construction, while builders might prioritize structural integrity and durability, clients may place a higher value on aesthetics, sustainability, or space efficiency. Understanding and aligning with these perspectives ensures that the final structure is not just solid but also a true reflection of the client's vision.

In Hardware Projects

Similarly, in hardware projects, while the team might focus on technical functionality and performance, the end-user's value might lie in ease of use, reliability, or compatibility with existing systems. Ensuring that the hardware delivers on these fronts guarantees not just functionality but also user satisfaction.

Continuous Feedback and Adaptation

Agile approaches champion a culture of continuous feedback and adaptation. This approach ensures that the project remains aligned with the customer's perspective throughout its lifecycle.

In Construction

In construction, this means regular client meetings, walkthroughs, and check-ins. It's about creating a feedback loop that ensures any misalignments are caught early and corrected, ensuring that the final structure aligns perfectly with the client's vision.

In Hardware Projects

In hardware projects, this involves engaging end-users in usability testing, gathering their feedback on system functionality, and iterating on the design based on this input. It ensures that the hardware not only meets but exceeds user expectations.

Conclusion: The Power of Alignment

In conclusion, understanding and aligning with the customer's perspective is a powerful force in ensuring project success. Whether in construction, hardware projects, or other seemingly linear fields, aligning project goals with customer needs and embracing a culture of continuous feedback are fundamental agile principles. This customer-centric approach ensures that the final product resonates with the end-users, fostering satisfaction and success.

In the chapters that follow, we will explore further applications of agility in different contexts, showcasing its universal adaptability and the profound impact it can have on project success, irrespective of the industry.

Chapter 8: The Customer's Perspective: Value from Their Eyes

1. **Understanding Customer Value**: Discuss strategies for understanding and prioritizing what is most valuable to the customer in an Agile project.
    - *Guidelines*: Focus on the importance of customer collaboration, empathy, and continuous feedback. Provide examples of how teams can ensure they are aligned with customer needs.
2. **Balancing Different Customer Needs**: Explore how teams can balance and prioritize the needs of different customers or stakeholders in a single project.
    - *Guidelines*: Discuss the challenges of managing diverse expectations and needs, providing strategies for effective communication, prioritization, and compromise.
3. **Customer Feedback in Agile**: How can teams effectively gather and respond to customer feedback in an Agile project, and what role does this play in delivering value?
    - *Guidelines*: Emphasize the importance of continuous feedback, adaptability, and customer collaboration. Provide examples of how teams can use feedback to enhance the value of their deliverables.

Chapter 9: Agile Challenges and Solutions: Thriving Amid Complexity

> "
>
> *"Real change, enduring change, happens one step at a time."*
>
> *– Ruth Bader Ginsburg (U.S. Supreme Court Justice)*

Introduction

Navigating the turbulent waters of agility in industries traditionally marked by linearity, such as construction and hardware projects, presents a unique labyrinth of challenges. This chapter is dedicated to dissecting these complexities, shedding light on the common hurdles encountered when infusing agile principles into these fields, and providing a beacon of practical solutions to navigate through them. In doing so, we aim to illuminate the path to agility, demonstrating its resilience and capacity to adapt and flourish, irrespective of industry constraints.

Breaking Work Down into Sprints or Iterations

One of the most significant challenges in applying agile to construction and IT hardware projects is the difficulty in breaking down large, complex tasks into smaller, manageable sprints or iterations.

Construction

In construction, projects are often large-scale and can take months or even years to complete. The traditional predictive or at best, waterfall model has been the standard approach, with each phase of the project being completed before the next one begins.

Solution: To overcome this challenge, it is crucial to adopt a more modular approach to construction. This could involve breaking the project down into smaller sections or phases, and completing each one before moving on to the next. Utilizing prefabrication and modular construction techniques can also help to make the construction process more agile.

IT Hardware and Networking

Similarly, in IT hardware and networking, projects can be complex and time-consuming, with a need for meticulous planning and execution.

Solution: Adopting a phased approach, where the network is built and implemented in stages, can help to make the project more manageable. This allows for testing and adjustments to be made at each stage, ensuring that any issues are identified and addressed early on.

Resistance to Change

The human psyche is wired to resist change, and this resistance can be a formidable barrier to adopting agile practices.

Solution: Cultivate a culture steeped in openness and transparency. Engage in open dialogues about the benefits of agility, and actively involve team members in the decision-making process, ensuring they feel a sense of ownership and commitment to the agile transformation.

Client Expectations

Clients accustomed to traditional project management approaches may have expectations misaligned with agile principles. Employ the use of information radiators like a Kanban board to clearly show what is being worked on. Use User Stories to clarify requested value and expectations.

Solution: Establish clear and consistent communication channels with clients. Regularly update them on project progress and educate them on the agile approach, ensuring they understand its benefits and how it aligns with their ultimate project goals.

Resource Constraints

Agile projects often require flexible resource allocation, which can be challenging in industries with rigid structures.

Solution: Prioritize tasks based on their value and urgency, ensuring that resources are allocated efficiently. Engage stakeholders in discussions about resource needs, and be prepared to adjust project timelines and expectations in response to resource availability.

Traditional Mindset

A traditional mindset, rooted in rigid processes and a resistance to change, can stifle agile adoption.

Solution: Invest in training and education to instill agile principles across the team. Showcase the tangible benefits of agility through practical application, turning skeptics into believers.

Lack of Documentation

Agile's emphasis on working products over comprehensive documentation can lead to concerns about insufficient project documentation.

Solution: Strike a balance between maintaining agility and ensuring that critical project information is documented and accessible. Avoid excessive paperwork, but ensure that essential documentation is not neglected.

Resistance to Iterative Approaches

Iterative approaches, a hallmark of Agile, can be a significant departure from traditional project management practices.

Solution: Introduce iterative practices gradually, allowing the team to experience their benefits firsthand. Foster a culture that embraces change and encourages active participation in the iterative process.

Scope Creep

The agile principle of responding to change can sometimes lead to scope creep, where the project's objectives continually expand.

Solution: Maintain a well-defined project scope, but remain open to necessary changes. Document and prioritize requested changes, involving the client in the decision-making process to ensure alignment with project goals.

Managing Dependencies

Agile projects, particularly in complex industries, often involve numerous dependencies that can impact project flow.

Solution: Proactively identify and track dependencies from the outset of the project. Utilize agile tools to manage and visualize dependencies, allowing for timely mitigation of potential issues.

Adaptability to Technological Changes

In industries marked by rapid technological advancements, the ability to adapt to new technologies is crucial.

Solution: Stay abreast of industry trends and emerging technologies. Build flexibility into project plans to accommodate technological changes, ensuring the project remains relevant and competitive.

Conclusion: Embracing Challenges as Opportunities

In the crucible of complexity, agility finds its true test. This chapter has navigated through the labyrinth of challenges, providing practical solutions and shining a light on the path to agile success in seemingly linear industries. By embracing these challenges as opportunities for growth and transformation, professionals can unlock the full potential of agility, turning hurdles into stepping stones towards excellence. As we venture into the next chapter, we venture deeper into the realms of continuous innovation and adaptation, underscoring the indispensable value of agility in ever-evolving environments.

Chapter 9: Agile Challenges and Solutions: Thriving Amid Complexity

1. **Navigating Complexity with Agile**: Discuss how Agile practices can help teams navigate and thrive in complex project environments.
    - *Guidelines*: Focus on the adaptability, transparency, and continuous improvement aspects of Agile, providing examples of how these can enhance project success in complex situations.
2. **Overcoming Common Agile Challenges**: Identify common challenges faced by Agile teams and discuss strategies for overcoming them.
    - *Guidelines*: Encourage problem-solving and critical thinking, providing examples of common Agile challenges and practical solutions.
3. **Agile and Risk Management**: Explore the role of Agile practices in effective risk management, discussing how teams can proactively identify and mitigate risks.
    - *Guidelines*: Highlight the importance of transparency, continuous inspection, and adaptability in managing risks in an Agile project.

Chapter 10: Continuous Innovation and Adaptation: The Agile Imperative

"Innovation is a bottoms-up, decentralized, and unpredictable thing, but that doesn't mean it cannot be managed."
– Eric Ries (Author, The Lean Startup)

Introduction

In today's fast-paced and ever-evolving landscape, the ability to innovate and adapt swiftly stands as a cornerstone for success and sustainability. This chapter meticulously unravels the critical importance of embedding continuous innovation and adaptation within the realms of construction, hardware, and other industries traditionally perceived as linear. We venture deep into the ethos of agility, illustrating that it transcends beyond the mere completion of projects, evolving into a catalyst for fostering a vibrant culture of innovation and ensuring organizations stay ahead of the curve.

The Need for Continuous Innovation

Changing Landscape

- **Construction**: The industry is witnessing a paradigm shift with the advent of new building materials, innovative construction techniques, and a growing emphasis on sustainability and eco-friendly practices.
- **Hardware Projects**: There is a relentless pace of technological advancements, coupled with rapidly evolving user needs and expectations, necessitating a constant state of innovation.

Competitive Pressure

- **Construction**: The competitive landscape demands innovative design solutions, integration of sustainable practices, and efficient construction methodologies to stay ahead.
- **Hardware Projects**: Staying relevant in this domain requires a proactive approach to embracing emerging technologies, understanding market dynamics, and adapting to them swiftly.

Agile as a Catalyst for Innovation

Iterative Experimentation

- **Construction**: Employing Agile, facilitates iterative experimentation with sustainable materials and construction techniques, enabling the discovery of optimal solutions.
- **Hardware Projects**: Agile practices encourage testing and iterating on hardware solutions, ensuring functionality, performance, and user experience are continually enhanced.

Customer Feedback Loops

- **Construction**: Establishing robust feedback loops with clients and end-users ensures that building designs and functionalities are continually refined and aligned with expectations.
- **Hardware Projects**: Agile practices underscore the importance of incorporating user feedback directly into the hardware design process, resulting in solutions that are truly user-centric.

Cross-Functional Collaboration

- **Construction**: Agile fosters a collaborative environment, bringing together architects, engineers, and contractors to innovate and solve problems collectively.
- **Hardware Projects**: It encourages the formation of cross-functional teams, blending hardware engineering, software development, and UX design to create holistic and integrated solutions.

Agile as a Driver of Adaptation

Flexibility in Planning

- **Construction**: Agile practices provide the flexibility needed to adapt construction plans in response to unforeseen challenges, changing site conditions, or evolving client requirements.
- **Hardware Projects**: They ensure that network configurations and hardware setups remain agile, capable of adapting to the changing technology landscape.

Risk Mitigation

- **Construction**: Agile practices play a pivotal role in proactively identifying and addressing potential construction risks, ranging from weather-related delays to material availability.
- **Hardware Projects**: They emphasize the importance of staying vigilant about cybersecurity threats and adapting security measures proactively to mitigate risks.

Scalability and Modularity

- **Construction**: Designing with scalability and modularity in mind ensures that buildings can accommodate future expansions or modifications without significant overhauls.
- **Hardware Projects**: Developing hardware systems that are scalable and modular ensures seamless integration with new technologies and components as they emerge.

Fostering a Culture of Agility and Innovation

Leadership Commitment

- **Construction**: Leaders play a crucial role in embedding agile practices within teams, encouraging innovation, and fostering a culture that thrives on continuous improvement.

- **Hardware Projects**: Project managers are empowered to prioritize adaptability, championing a culture that is resilient, innovative, and perpetually evolving.

Education and Training

- **Construction**: Providing ongoing education and training on sustainable construction practices, innovative materials, and new methodologies ensures teams are well-equipped to innovate. Cross-training and job-shadowing also enhances agility in any industry by enabling the team to synergize and collaborate more effectively and seamlessly.
- **Hardware Projects**: Ensuring that team members are constantly updated with the latest technological advancements and security protocols keeps the organization at the forefront of innovation. Business leaders should intentionally propose cross-training, pair-programing and job-shadowing as part of each team-members perpetual goals.

Continuous Integration and Testing in Hardware Development

In the realms of construction and engineering, hardware development traditionally stands as a distinct phase, often culminating in a bottleneck of integration and testing activities towards the project's conclusion. This sequential approach, while established, is prone to inefficiencies, leading to potential delays, the necessity for rework, and escalated costs.

Embracing Agile philosophies offers a transformative solution, encouraging professionals to embed continuous integration and testing throughout the hardware development lifecycle. This shift fosters a culture of early issue detection, rapid feedback, and enhanced collaborative synergy among all project participants.

Specifically, in hardware development, continuous integration translates to the ongoing amalgamation of hardware components and subsystems. This proactive approach ensures timely identification and resolution of compatibility issues or conflicts, mitigating the risk of substantial project delays or setbacks.

Collaborative Design in Agile Construction

Collaborative design stands as a cornerstone of Agile construction, fostering a harmonious environment where professionals from diverse fields—engineers, architects, project managers, and other stakeholders—converge to innovate and solve problems collectively.

In the fast-paced and ever-evolving landscape of the construction industry, the capacity to swiftly adapt and cater to client needs is paramount. Agile project management methodologies have demonstrated their efficacy in meeting these demands, with collaborative design playing an instrumental role in their successful deployment.

This collaborative ethos encourages open dialogue and active participation from all team members, ensuring a holistic integration of various disciplines and expertise. Engaging stakeholders from the project's inception allows the team to tap into a wealth of collective knowledge, proactively addressing potential challenges, fostering innovation, and making well-informed decisions.

For practitioners in hardware and engineering, the Agile construction framework, enriched by collaborative design, offers a plethora of advantages. It supports iterative development and prototyping, establishing rapid feedback loops and facilitating real-time design testing and refinement. This iterative process not only accelerates development but also guarantees that the final product aligns precisely with client specifications.

In the specific context of linear infrastructure projects—roads, railways, pipelines, and the like—collaborative design is instrumental in streamlining both planning and execution. Engaging a diverse group of stakeholders, from project managers to engineers, ensures the Agile philosophy is applied effectively, optimizing project delivery, curtailing costs, and boosting overall efficiency.

Extending its influence beyond the construction phase, collaborative design also plays a vital role in supply chain management and manufacturing processes within hardware development. Here, Agile principles empower organizations to swiftly respond to market fluctuations, minimize waste, and enhance product quality, ensuring a competitive edge in a demanding industry.

Conclusion: The Agile Imperative

In wrapping up this chapter, it becomes unequivocally clear that continuous innovation and adaptation are not just beneficial but imperative for thriving in the construction, hardware, and other traditionally linear industries. Embracing agile principles is a strategic move, empowering professionals and organizations alike to navigate through evolving landscapes, respond adeptly to changing customer needs, and maintain a competitive edge. This chapter serves as a testament to the transformative power of agility, underscoring its role as a vital catalyst for fostering a culture rich in innovation and poised for long-term success.

Chapter 10: Continuous Innovation and Adaptation: The Agile Imperative

1. **Fostering a Culture of Innovation**: Discuss how Agile practices can contribute to creating a culture of continuous innovation within a team or organization.
    - *Guidelines*: Focus on the empowerment, collaboration, and continuous learning aspects of Agile, providing examples of how these can foster innovation.
2. **Agile and Organizational Change**: Explore how adopting Agile practices can drive positive organizational change, discussing potential challenges and strategies for success.
    - *Guidelines*: Discuss the transformative power of Agile, providing examples of how it can lead to more adaptive, resilient, and customer-centric organizations.
3. **Sustaining Agility Over Time**: How can teams and organizations ensure that they sustain their Agile practices and continue to innovate and adapt over time?
    - *Guidelines*: Emphasize the importance of continuous improvement, learning, and staying true to Agile principles. Provide strategies for maintaining momentum and avoiding complacency.

Chapter 11: Agile Success Stories: Real-World Examples of Transformation

> *"Plans are worthless, but planning is everything."*
>
> *– Dwight Eisenhower (General and President)*

Introduction

Agility, often misconstrued as a concept exclusive to software development, has proven its transformative power across various industries, including the seemingly linear realms of construction and hardware projects. This chapter ventures into compelling real-world examples, illustrating how professionals have navigated challenges, delivered outstanding results, and fostered innovation, all under the agile umbrella.

CONSTRUCTION SOFTWARE - CASE STUDY A

Summary of "A Six-Month Cultural Transformation - The Penta Story"

Penta Technologies, a family-owned construction software company founded in 1991 in Milwaukee, WI, faced a significant challenge. The company operated in a traditional construction company manner, leading to a complex and hard-to-scale product due to customer-driven functionality. This resulted in a slow and inconsistent delivery pace, employee burnout, and a lack of innovation.

Recognizing the need for change, the company brought in new leadership in 2018, including Bill Wagner as President and Laura Henderson as COO, both with a history of transforming software businesses. They initiated a cultural shift, starting with identifying core problems and waste, and eventually transitioning to the Scrum framework for a more agile and efficient operation.

The transition was not without its challenges. The company faced financial pressures as revenue from customer work decreased, and there was a need for greater customer validation and technical excellence. However, through transparency, empowerment, and a focus on delivering value, the company was able to create a more positive work environment, increase employee empowerment, and improve product delivery.

The results of the transformation were clear. Employee negativity decreased, empowerment increased, and the company was able to consistently deliver "Done" increments every Sprint. The focus shifted from top-down mandates to collaborative problem-solving, data-informed decision-making, and a culture of trust and transparency.

In summary, Penta Technologies underwent a significant agile transformation, moving from a traditional and inefficient operation to a more agile, efficient, and employee-centric organization. The leadership's commitment to change, focus on identifying and eliminating waste, and adoption of the Scrum framework were key to this successful transformation.

OIL AND GAS - CASE STUDY B

Summary of "The Journey to Scrum@Scale at a Fortune Oil & Gas Company"

Introduction

Scrum Inc. played a crucial role in transforming a Fortune 500 Oil & Gas Company's approach to project management and execution. The company faced challenges due to a growing IT portfolio from numerous acquisitions, leading to a lack of integration among 150 legacy IT systems. This case study outlines Scrum Inc.'s unique approach to implementing Scrum@Scale to address these challenges.

Key Differentiators of Scrum Inc.'s Approach

1. **Minimal Viable Planning**: Rapid pilot deployment through just enough planning.
2. **Leveraging Past Client Experience**: Utilizing experience to quickly identify and manage risks.
3. **Engaging the C-Suite and Addressing Cultural Issues**: Proactively eliminating cultural hurdles for faster progress.
4. **Learning from Failures**: Analyzing past and current failures to inform future steps.

Background and Challenges

The company decided to implement an SAP ERP system to streamline its IT infrastructure. Despite a consistent investment of $50 million per month, progress was hard to demonstrate, leading to the exploration of Scrum as a potential solution.

Agile Transformation with Scrum@Scale

Scrum Inc. trained the executive team and helped form executive coordination teams and Scrum Key Performance Objectives (KPOs). The company then engaged Scrum Inc. to train and coach the SAP implementation, enterprise solutions, and data wrangling teams in Scrum practices.

Case in Focus: Wells and Fields Data Team

The Wells and Fields Data Team faced challenges such as geographically distributed teams, limited SME resources, and insufficient development staff. With Scrum Inc.'s guidance, they adopted one-week Sprints, measured team happiness, addressed overcommitment issues, and experimented with co-location. This resulted in a significant increase in productivity and team morale.

Enterprise Scrum Engagement and Success Stories

- **Executive Team**: Training of 50 executives, organization of Scrum teams, and establishment of Scrum KPOs.
- **SAP Implementation Team**: Reduction in contractors, training of 37 teams, and significant progress by mid-Q2 2018.
- **Field Logistics Management Tool Team**: Delivery of a Field Logistics Management Tool integrated with SAP over 8 months, with high stakeholder engagement.

Results

- The Wells and Fields Data Team delivered North American Wells and Fields data into the new ERP system one week ahead of schedule, with a 93% readiness level.
- Readiness levels for two other mission-critical countries were significantly improved.
- The Field Logistics Management Tool Team delivered a highly relevant tool, with a 2x improvement in output and 20% increase in productivity.

Conclusion

Scrum Inc.'s approach to implementing Scrum@Scale at this Fortune 500 Oil & Gas Company resulted in a transformative change, overcoming integration challenges, improving productivity, and delivering tangible results. The company's journey from a state of disarray to agile efficiency showcases the power of Scrum and Scrum@Scale in driving organizational change and project success.

AERONAUTICS - CASE STUDY C

Summary of "Owning the Sky with Agile" (Saab Aeronautics Story)

This is a case study by Jörgen Furuhjelm, Johan Segertoft, Joe Justice, and J.J. Sutherland, detailing how Saab Aeronautics utilized Agile practices to develop the JAS 39E Saab Gripen, a multi-role strike fighter jet. The study highlights the escalating costs of defense procurement and how Agile practices were implemented to control costs while delivering high-quality results.

Key Points:

1. **Background**: The cost of defense procurement has been rising significantly, prompting companies like Saab Defense to seek innovative ways to control costs. Saab Aeronautics adopted Agile practices for both hardware and software teams in the development of the Gripen E fighter jet.
2. **Agile Implementation**: Agile practices were introduced at Saab approximately ten years ago, starting with small software development teams. The practices have since evolved and spread across the organization, encompassing software, hardware, and fuselage design. The Saab Agile framework includes practices from Lean, Scrum, Kanban, XP, and others, allowing teams the autonomy to tailor the framework to their specific needs.
3. **Commitment and Clarity**: Saab emphasizes the importance of commitment at the individual and team levels, as well as clarity in understanding the project's goals. Decisions are driven down to the team level, minimizing organizational hierarchy and promoting transparency.
4. **Managing Variability**: The development of the Gripen E involves over 1000 engineers and more than 100 teams. Agile practices, particularly Scrum, are used to manage the inherent variability in such a complex project, with a stable pulse and common rhythm established across all teams.
5. **Prioritization**: Regular cycles of work, including three-week Sprints and quarterly Increments, help establish a rhythm for development. Prioritization of the product backlog is crucial to ensure that teams are working on the most valuable tasks, and this is facilitated through the role of the Product Owner.
6. **Strategic Planning**: While Agile practices are central to Saab's approach, strategic planning remains a vital component. The strategic plan acts as a constraint, helping teams visualize their roadmap and align their work with the overall project goals.

7. **Continuous Improvement**: Continuous improvement is embedded in the Agile practices at Saab, with regular Retrospectives held to identify areas for improvement. This approach extends across teams, with a Retrospective of Retrospectives conducted to address common issues and share learnings.
8. **Agile Catalysts**: The modular design of the Gripen E, the use of state-of-the-art simulators, and the close interaction between development teams and pilots are highlighted as key factors enabling Agile practices at Saab.
9. **Conclusion**: The implementation of Agile practices at Saab Aeronautics has resulted in the development of the Gripen E fighter jet with lower costs, faster delivery, and higher quality. The study underscores the importance of managing variability, driving performance with clarity and commitment, and continuously improving processes to achieve these outcomes.

The document serves as a comprehensive case study on the successful application of Agile practices and philosophies in a highly complex and traditionally rigid field, showcasing the potential for Agile to drive efficiency, innovation, and quality in large-scale defense projects.

BIOTECH - CASE STUDY D

Title: Leveraging Scrum in Biotech: A Case Study on Dyno Therapeutics

Executive Summary:

This case study explores the innovative application of Scrum within the biotech industry, focusing on Dyno Therapeutics, a pioneering organization in gene therapy. Tyson Bertmaring, Head of Partner Success at Dyno, shares his journey and experiences in implementing Scrum, shedding light on how this Agile framework has facilitated a collaborative, transparent, and adaptive work environment. Hosted by Dave West, this episode of the Scrum.org community podcast provides valuable insights for organizations looking to enhance their agility and efficiency in complex research domains.

1. **Introduction**

1.1 **Background**: Dyno Therapeutics stands at the forefront of gene therapy, utilizing artificial intelligence to revolutionize treatment methods. The organization's commitment to research and development necessitates a flexible and responsive approach to project management.

1.2 **Objective**: This case study aims to dissect the implementation of Scrum at Dyno, understanding its impact on the organization's workflow, culture, and overall success in the biotech field.

2. **The Journey to Scrum**

2.1 **Initial Exposure**: Tyson's first encounter with Scrum was during his tenure in the military, where traditional project management was the norm. The rigid structure of traditional methods proved to be a misfit for the dynamic nature of biotech research, prompting a shift towards Agile practices.

2.2 **Adoption at Dyno**: Joining Dyno Therapeutics, Tyson saw the potential for Scrum to thrive in an environment that valued learning, adaptation, and results. The leadership's openness to Agile practices laid a strong foundation for Scrum's integration.

3. **Implementing Scrum in Biotech**

3.1 **Challenging Skill Silos**: The biotech industry is characterized by highly specialized skills and expertise. Scrum at Dyno facilitated a shift from a skill-centric to a goal-centric approach, fostering cross-functional collaboration and enhancing the team's adaptability.

3.2 **Building a Learning Organization**: The iterative nature of Scrum, coupled with its emphasis on transparency and inspection, resonated with Dyno's commitment to continuous learning and improvement. This alignment proved crucial in navigating the complexities of gene therapy research.

4. **Outcomes and Benefits**

4.1 **Enhanced Transparency and Feedback**: The adoption of Scrum led to increased transparency across teams, fostering an environment where feedback was readily exchanged, and ideas were continuously refined.

4.2 **Fostering a Value-Driven Culture**: The shift towards a goal-centric approach cultivated a culture where the value of work took precedence over individual skills, driving the team towards common objectives and shared success.

5. **Reflections and Future Prospects**

5.1 **The Journey Continues**: Two years into their Agile transformation, Dyno acknowledges that there is still much to learn and adapt. The journey thus far, however, has laid a robust foundation for future success.

5.2 **A Model for Biotech**: Dyno's experience serves as a testament to the viability of Scrum in biotech, providing a blueprint for other organizations in the field to follow.

6. **Conclusion**

This case study encapsulates the transformative journey of Dyno Therapeutics, illustrating how Scrum can be effectively applied in the biotech industry to foster innovation, collaboration, and efficiency. The insights shared by Tyson Bertmaring highlight the pivotal role of Agile approaches in navigating the complexities of gene therapy research, setting the stage for continued growth and success in the field.

AEROSPACE, DEFENSE & SECURITY, TRANSPORTATION - CASE STUDY E

Title: Scaling Scrum @ Thales with Emergent Innovation

Summary: The case study revolves around Thales, a company operating in various markets including aerospace, space, ground transportation, defense, and security. The focus is on the Technical Unit Processing (TU Processing) within the Surface Radar business line in The Netherlands, which is responsible for developing digital processing functions for naval and air defense applications.

1. **Background**: TU Processing, consisting of over 70 people, decided to experiment with Agile and the Scrum Framework about ten years ago, despite initial doubts about its applicability in a high-tech embedded software development environment with defense requirements and minimal end-customer contact.
2. **Initial Agile Adoption**: The department started experimenting with Agile using student projects, research projects, and later, small development projects in the radar domain. The initial trials showed promising results, leading to a decision to take the next step in Agile adoption.
3. **Need for Change**: The long duration of projects, the need for specialized knowledge in various subdomains, and increasing market demands for faster delivery without compromising quality and reliability necessitated a change in the working methodology.
4. **Adoption of Scrum**: The department decided to roll out Scrum across the entire department with the main objective of maintaining reliability while improving development success through increased transparency, control, and focus on value.
5. **Transition to Stable Teams**: The first major change was the transition from project teams to stable, long-lasting cross-functional Scrum teams. This change aimed to enhance knowledge sharing, reduce dependency on individual specialists, and improve overall productivity.
6. **Challenges and Moderate Results**: Despite increased project transparency and positive feedback from engineers, there were challenges such as disturbances from system testers, lack of stable velocity, and dissatisfaction among stakeholders, indicating deeper issues that needed addressing.
7. **Emergent Innovation and Scaling Scrum**: The department engaged with AgiliX and adopted the Emergent Innovation framework to scale Scrum at a larger level. This involved

forming a leadership change team, creating an initial innovation plan, communicating the change, and learning with Optimization Teams.

8. **Results and Future Steps**: The adoption of Agile and Scrum led to clearer and more reliable steering information for stakeholders, reduced dependency on key individuals, increased focus on high-priority work, and improved quality of work. Future steps include introducing team-based incentives and organizing more around customer value.

The case study highlights the journey of TU Processing at Thales in adopting and scaling Scrum, showcasing the challenges, strategies, and results of this transformative process.

HARDWARE: NETWORK INSTALLATION REINVENTED HYPOTHETICAL CASE STUDY

Success Story

A leading tech company faced the daunting task of implementing a comprehensive network installation across a sprawling corporate campus, encompassing multiple office buildings.

Agile Approach

- **Cross-Functional Teams**: The project was spearheaded by a diverse team of hardware engineers, software developers, and UX designers, all working in tandem to create a network that was both robust and user-friendly.
- **Iterative Sprints**: The installation was deconstructed into manageable sprints, each focusing on specific segments of the network or individual buildings.
- **User Feedback**: A continuous feedback loop was established with the end-users, ensuring that network configurations were tweaked and optimized for the best possible user experience.
- **Scalability**: The hardware components were meticulously chosen for their scalability, ensuring the network could effortlessly accommodate future technological advancements.

Outcome

The network installation was not only completed ahead of schedule but also exceeded user expectations in terms of reliability and performance, ensuring the company was well-prepared for future tech upgrades.

MANUFACTURING: LEAN AND AGILE PRODUCTION HYPOTHETICAL CASE STUDY

Success Story

A manufacturing giant decided to infuse agile principles into its production processes, aiming to enhance efficiency, product quality, and customer satisfaction.

Agile Approach

- **Value Stream Mapping**: The team engaged in a thorough analysis of the production processes, identifying and eliminating waste to optimize efficiency.
- **Daily Stand-Ups**: Daily meetings were instituted, providing a platform for team members to discuss challenges, share updates, and collaborate effectively.
- **Kanban Systems**: Visual boards were utilized to manage work in progress, ensuring a smooth and uninterrupted production flow.
- **Customer-Centricity**: The production processes were realigned to be in sync with customer demand, facilitated by regular feedback and the ability to quickly adapt to changing needs.

Outcome

The transformation led to a significant reduction in production lead times, a noticeable improvement in product quality, and a surge in customer satisfaction, ultimately giving the company a substantial competitive advantage.

Conclusion: Inspiring Transformations

These stories serve as a testament to the versatility and transformative power of agile thinking, demonstrating its applicability and effectiveness beyond the realm of software development. Whether it's revolutionizing construction practices, redefining network installations, or streamlining manufacturing processes, embracing agile principles enables professionals to deliver exceptional results, adapt to change, and drive innovation. These real-world examples stand as beacons of inspiration, showcasing the profound impact of agility across diverse industries and paving the way for future transformations.

CONSTRUCTION HYPOTHETCAL CASE STUDY

The Sustainable Build

Success Story

A renowned construction firm embarked on an ambitious journey to create a state-of-the-art sustainable office complex, catering to the vision of an environmentally conscious client.

Agile Approach

- **Customer Collaboration**: Engaging in frequent, transparent discussions with the client ensured a deep understanding of their evolving sustainability goals and requirements.
- **Iterative Design**: The design team, comprising architects and engineers, engaged in a continuous cycle of experimentation with sustainable materials and innovative energy-efficient designs.
- **Adaptation**: The project plan remained fluid, adapting to incorporate the latest in sustainability technologies as they became available.
- **Documentation**: A balance was struck between agility and documentation, ensuring all sustainability standards were meticulously adhered to.

Outcome

The project culminated in an office complex that not only achieved the highest sustainability certifications but also became a magnet for eco-conscious businesses, solidifying the construction firm's reputation as a leader in green building.

Chapter 11: Agile Success Stories: Real-World Examples of Transformation

1. **Analyzing Agile Transformations**: Choose a real-world example of an Agile transformation, and discuss the key factors that contributed to its success.
    - *Guidelines*: Encourage research and critical analysis, highlighting the role of leadership, culture, and continuous learning in successful Agile transformations.
2. **Lessons from Agile Failures**: Explore a case where an Agile transformation did not go as planned, discussing what went wrong and what lessons can be learned.
    - *Guidelines*: Emphasize the importance of understanding Agile principles, providing adequate training, and fostering an Agile culture. Discuss how failures can be valuable learning opportunities.
3. **Agile in Non-Software Contexts**: Provide examples of successful Agile transformations in industries outside of software development, discussing the unique challenges and benefits.
    - *Guidelines*: Highlight the universality of Agile principles, providing examples of how they can be applied in various contexts to drive improvement and innovation. Use the Scrum.org case study repository as a resource for examples: https://www.scrum.org/resources/case-studies

Chapter 12: Final Thoughts

> *"Sometimes it can be really freeing to have identified the problem... Even when you're it."*
>
> *– Kristen Belcher Agile coach*

Embarking on the journey of implementing Agile in hardware projects requires a nuanced approach, considering the unique challenges and opportunities that hardware development presents. Here's an expanded view on the profound thoughts shared, aiming to provide a comprehensive roadmap to agility in the hardware context.

1. **Valuing Individuals and Interactions:**

- **Fostering Collaboration**: Encourage open communication and collaboration within the team and across different departments. Hardware projects often involve various disciplines; ensuring everyone is on the same page is crucial.
- **Building a Supportive Culture**: Cultivate a culture that values every team member's input, fostering a sense of ownership and accountability. This is vital in hardware projects where the cost of changes increases significantly as the project progresses.
- **Adapting to Change**: Be prepared to adapt to changes in requirements, technology, and market demands. Hardware development cycles are traditionally longer, making adaptability a key factor for success.

2. **Focusing on a Working Product:**

- **Iterative Prototyping**: Develop iterative prototypes to validate concepts and designs. This approach helps in identifying potential issues early in the development cycle, saving time and resources.
- **Balancing Perfection and Progress**: Strive for a balance between perfection and progress. In hardware development, the pursuit of perfection can lead to significant delays. Focus on creating a working product and iterate upon it.]

3. **Collaborating with the Customer:**

- **Continuous Feedback**: Establish channels for continuous customer feedback. Regular interactions with the customer can provide valuable insights, helping to align the product development with customer needs.
- **Setting Realistic Expectations**: Hardware development has its constraints and complexities. Engage with the customer to set realistic expectations regarding timelines, costs, and deliverables.

4. **Responding to Change:**

- **Embracing Flexibility**: Develop a mindset that embraces change. In the fast-evolving tech landscape, being able to pivot and adapt to new information or market demands is crucial.
- **Balancing Agility and Stability**: Find the right balance between agility and stability. While being agile is important, hardware development requires a certain level of stability and thorough validation.

5. **Prioritizing Customer Satisfaction:**

- **Demonstrating Progress**: Break down the product into smaller, manageable pieces that demonstrate progress. Even if a piece is not immediately usable, showing progress can build customer trust and satisfaction.
- **Valuing Prototypes and Visuals**: Utilize prototypes, wireframes, and burn charts to visually demonstrate progress at Sprint reviews. This practice not only provides value to stakeholders but also aligns the team and uncovers potential issues.

6. **Empowering the Team:**

- **Providing Resources and Support**: Ensure that the team has the necessary resources and support to be agile. This is particularly important in hardware development where access to tools, components, and testing facilities is crucial.
- **Fostering Autonomy**: Encourage a culture of autonomy, allowing the team to make decisions and pivot without being overly reliant on a central figure like a project manager.

7. **Implementing Agile Practices:**

- **Utilizing Agile Artifacts and Events**: Adopt Agile practices such as a living product backlog, locked Sprint backlogs, regular backlog refinement, daily stand-ups, Sprint reviews, and retrospectives. These practices help in maintaining alignment, transparency, and continuous improvement.

- **Targeting Shippable Increments**: Define and target potentially shippable increments or meaningful increments at the end of each iteration, even in a hardware context where shipping might not be as frequent.

8. **Embracing a Hybrid Approach:**

- **Being Open to Hybrid Models**: Recognize that in some instances, a hybrid approach incorporating elements of predictive planning might be necessary. This is especially true in hardware development where certain components or processes might have long lead times or require upfront planning.
- **Experimenting and Adapting**: Encourage experimentation and be open to adapting the approach based on what works best for the team and the project. Avoid being overly prescriptive in the application of Agile practices.

9. **Understanding and Evolving:**

- **Mastering the Basics**: Initially, there might be a need to be directive and adhere closely to Agile frameworks such as Scrum or Kanban. Ensure that the team has a strong understanding of the rules, values, and principles of Agile.
- **Evolving with Maturity**: As the team matures in its Agile journey, foster an environment that encourages experimentation and deviation from rigid adherence, always keeping the core values and principles in mind.

Implementing Agile in hardware development is a challenging yet rewarding endeavor. By valuing individuals, focusing on delivering a working product, collaborating closely with customers, being ready to respond to change, prioritizing customer satisfaction, empowering the team, implementing Agile practices, embracing a hybrid approach, and understanding and evolving with Agile principles, teams can navigate the complexities of hardware development while reaping the benefits of agility.

Conclusion

This book has ventured deep into the intricacies of implementing Agile approaches in hardware projects, providing a comprehensive guide for those looking to navigate this challenging yet rewarding landscape. We have uncovered that Agile, while traditionally associated with software development, holds immense potential in transforming hardware projects, fostering a culture of collaboration, innovation, and continuous improvement.

The future belongs to those bold and visionary project managers who are willing to break the mold and apply Agile principles to hardware projects. Agile is not a passing trend; it is a robust and flexible approach that has stood the test of time, proving its value across various industries and project types.

As we stand at the crossroads of tradition and innovation, the choice is clear. Ignoring the call to Agile or finding reasons to stick to outdated and rigid project management methodologies will only lead to stagnation. The Agile journey requires courage, commitment, and a willingness to embrace change, but the rewards are well worth the effort.

I call upon you, the reader, to take this knowledge and use it as a springboard to propel your projects and teams forward. Embrace Agile with an open mind and a determined spirit, and watch as your hardware projects transform into models of efficiency, innovation, and success. The time to act is now. Step into the Agile future, and lead your projects to unparalleled heights of success and satisfaction. All the best in your Agile journey!

Your Friend Phill

References

1. Agile Manifesto
    - https://agilemanifesto.org
2. Company Spotlight - Scrum at Gillette
    - https://www.scrum.org/resources/company-spotlight-scrum-gillette
3. Company Spotlight - Dyno Therapeutics - Using Scrum in BioTech
    - https://www.scrum.org/resources/company-spotlight-dyno-therapeutics-using-scrum-biotech
4. Straçusser, G. (2015). Agile project management concepts applied to construction and other non-IT fields. Paper presented at PMI® Global Congress 2015—North America, Orlando, FL. Newtown Square, PA: Project Management Institute
    - https://www.pmi.org/learning/library/agile-software-applied-to-construction-9931
5. A Six-Month Cultural Transformation - The Penta Story
    - https://www.scrum.org/resources/six-month-cultural-transformation-penta-story
6. Scaling Scrum at Thales with Emergent Innovation
    - https://scrumorg-website-prod.s3.amazonaws.com/drupal/2016-06/Scrum.org-White-Paper-Thales.pdf
7. 101+ Inspiring Quotes About Agile by Mountain Goat Software
8. The Journey to Scrum@Scale at a Fortune 500 Oil & Gas Company
    - https://www.scruminc.com/journey-to-scrum-at-scale-fortune-500-oil-gas-company

About the Author

Phill C. Akinwale, PMP, is a highly experienced project management professional with a proven track record in both government and private sectors. Throughout his career, he has worked with renowned companies such as Motorola, Honeywell, Emerson, Skillsoft, Citigroup, Iron Mountain, Brown and Caldwell, US Airways, and CVS Caremark, managing operational endeavors, projects, and project controls.

With his extensive knowledge and expertise in various facets of Project Management, Phill has become a trusted figure in the field. He has trained project management professionals worldwide, including prestigious organizations like NASA, FBI, USAF, USACE, US Army, and the Department of Transport. Over the past 15 years, he has provided training across five editions of the PMBOK® Guide, showcasing his commitment to staying up-to-date with industry standards.

Originally from a Civil Engineering (BEng) and Construction Information Technology (M.Sc.) background, Phill holds an impressive twelve project management certifications, with a notable focus on Agile Project Management. His certifications include Certified ScrumMaster (CSM), PMI Agile Certified Practitioner (PMI-ACP), Professional Scrum Master (PSM), Professional Scrum Product Owner (PSPO), Professional Agile Leadership (PAL), and Scaled Professional Scrum (SPS). This diverse range of certifications reflects his dedication to mastering various project management methodologies.

In addition to his project management expertise, Phill is a certified coach and speaker through the John Maxwell team. Leveraging his leadership and soft skills, he delivers impactful workshops, seminars, keynote speeches, and coaching sessions. He is passionate about guiding individuals, teams, and organizations in their desired direction, equipping them with the tools and knowledge to achieve their goals.

Phill is also a prolific author, having written over 20 books on a diverse range of topics. His written works cover areas such as bullying (including "The Bird Brained Bullies"), leadership, conflict resolution, time travel ("The Time Machine Project"), and even comics ("Project VBX 11"). Furthermore, he has created two short movies that explore themes of leadership drama in the workplace and project management.

With his vast experience, certifications, and passion for guiding others, Phill C. Akinwale, PMP, is a valuable asset to any organization or individual seeking agile coaching, project management expertise, leadership development, and personal growth.

Made in the USA
Monee, IL
05 November 2026

67812870R00076